T0257792

Handbook of Fiber Gratings

Handbook of Fiber Gratings

Edited by **Bridget Crafts**

New York

Published by NY Research Press,
23 West, 55th Street, Suite 816,
New York, NY 10019, USA
www.nyresearchpress.com

Handbook of Fiber Gratings
Edited by Bridget Crafts

International Standard Book Number: 978-1-63238-249-8 (Hardback)

Printed in the United States of America.

Contents

Preface

Every book is initially just a concept; it takes months of research and hard work to give it the final shape in which the readers receive it. In its early stages, this book also went through rigorous reviewing. The notable contributions made by experts from across the globe were first molded into patterned chapters and then arranged in a sensibly sequential manner to bring out the best results.

This book includes of a compilation of valuable information contributed by distinct veterans from across the globe, elucidating the recent research trends in both short- and long-period fiber grating technology. It is primarily intended for researchers already engaged in this area, but it is also useful for anybody with a scientific background who is interested in an up-to-date analysis of the current growth in this domain. It will also prove to be useful for engineers and scientists who have become recently involved in this field. The aim of this book is to examine the present research in this area, displaying the current progresses in the field of fiber gratings.

It has been my immense pleasure to be a part of this project and to contribute my years of learning in such a meaningful form. I would like to take this opportunity to thank all the people who have been associated with the completion of this book at any step.

Editor

A Guide to Fiber Bragg Grating Sensors

Marcelo M. Werneck, Regina C. S. B. Allil, Bessie A. Ribeiro
and Fábio V. B. de Nazaré

Additional information is available at the end of the chapter

1. Introduction

Optical fiber sensors (OFS) appeared just after the invention of the practical optical fiber by Corning Glass Works in 1970, now Corning Incorporated, that produced the first fiber with losses below 20 dB/km. At the beginning of this era, optical devices such as laser, photodetectors and the optical fibers were very expensive, afforded only by telecom companies to circumvent the old saturated copper telephone network. With the great diffusion of the optical fiber technology during the 1980's and on, optoelectronic devices became less expensive, what favored their use in OFS.

OFS can be applied in many branches of the industry but we will concentrate here the electrical power industry. In this area, the operators need to measure and monitor some important physical parameters that include:

- Strain ($\mu\epsilon$)
- Vibration of structures and machines
- Electric current (from A to kA)
- Voltage (from mV to MV)
- Impedance ($\mu\Omega$)
- Leakage current of insulators (μA to mA)
- Temperature
- Pressure
- Gas concentration
- Distance between stationary and rotating or moving parts

In the electrical power industry (EPI) we have two facts that can cause collapse of an electronic sensor: presence of high voltage and presence of high electromagnetic interference. Therefore, depending on where we want to measure a parameter it can be very difficult or even impossible to use a conventional sensor. The best option to circumvent this is through the use of an OFS, because the fiber is made of dielectric material sand, therefore, it is possible to place them very close or even over a high potential conductor and they do not necessarily need electrical power at the sensor location.

Another problem with conventional sensors is that they all need electric energy to work. However, providing electric energy at the sensor location is sometimes difficult if the device needs to be far away from any appropriated power supply. It happens in long high voltage transmission lines, at high voltage potentials, along pipe-lines or in deep ocean, for instance. Since OFS are passive sensors they do not need electric energy to work.

Therefore we can mention some, very specific characteristics of OFS that are well exploited when applied to the EPI:

- High immunity to Electromagnetic Interference (EMI)
- Electrical insulation
- Absence of metallic parts
- Local electrical power not required
- Lightweight and compactness
- Easy maintenance
- Chemically inert even against corrosion
- Work over long distances
- Several sensors can be multiplexed on the same fiber

There are many options to develop an OFS. The easiest way is by making the measurement to modulate the light amplitude that is the power, and ending up with an amplitude modulated sensor. These sensors were very common at the beginning of OFS era but they gradually were substituted by wavelength based sensors. These are more stable and self-calibrated as the wavelength does not depend on losses due connectors, modal drifts, macro bends, or LED and LASER ageing/drifts.

In this Chapter we will concentrate on a very special type of OFS: the Fiber Bragg Grating (FBG) sensors.

2. Theory and models of FBG

Fiber Bragg Grating (FBG) technology is one of the most popular choices for optical fiber sensors for strain or temperature measurements due to their simple manufacture, as we will see later on, and due to the relatively strong reflected signal. They are formed by a periodic modulations of the index of refraction of the fiber core along the longitudinal direction and can be produced by various techniques. The term fiber Bragg grating was borrowed from the Bragg law and applied to the periodical structures inscribed inside the core of conventional telecom fiber. Therefore, before entering the theory of fiber Bragg grating itself, it is worth to go back one century behind in order to review the Bragg law.

Sir William Lawrence Bragg, was born in 1890, a British physicist and X-ray crystallographer, was the discoverer, in 1912, of the Bragg law of X-ray diffraction. This principle is used until today for the study and determination of crystal structure, particularly in thin film research. Sir Bragg, together with his father, won the Nobel Prize for Physics in 1915 for an important step in the development of X-ray crystallography.

Bragg diffraction occurs for an electromagnetic radiation whose wavelength is the same order of magnitude of the atomic spacing, when incident upon a crystalline material. In this case the radiation is scattered in a specular fashion by the atoms of the material and experiences constructive interference in accordance to Bragg's law. For a crystalline solid with lattice planes separated by a distance d the waves are scattered and interfere constructively if the path length of each wave is equal to an integer multiple of the wavelength. Figure 1 shows the idea. Bragg's law describes the condition for constructive interference from several crystallographic planes of the crystalline lattice separated by a distance d:

$$2d \sin\theta = n\lambda \tag{1}$$

Where θ is the incident angle, n is an integer and λ is the wavelength. A diffraction pattern is obtained by measuring the intensity of the scattered radiation as a function of the angle θ. Whenever the scattered waves satisfy the Bragg condition it is observed a strong intensity in the diffraction pattern, known as Bragg peak.

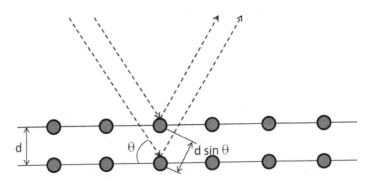

Figure 1. An incident radiation is reflected by the lattice structure of a crystal and will interfere constructively if the Bragg law is obeyed.

The first observations of index of refraction changes were noticed in germane silicate fibers and were reported by Kenneth Hill and co-workers in 1978 [1]. They described a permanent grating written in the core of the fiber by an argon ion laser line at 488 nm launched into the fiber by a microscope objective. This particular grating had a very weak index modulation, resulting in a narrow-band reflection filter at the writing wavelength. In reality, this phenomenon happened by chance, when they injected a high power blue light into de fiber, unexpectedly, after a few minutes, the transmitted light decayed. It was Friday but they were so puzzled with this phenomenon that Hill returned to his laboratory on Saturday to make a new experiment. He wanted to know where the light was going to and he had a clue. He used a thin microscope slide as a beam splitter in order to monitor a possible reflection from the fiber and there was the missing light [2]. The explanation is that at the end of the fiber about 4% of the light was reflected by Fresnel reflection which, in its way backwards, interfered with the ongoing light producing an interference pattern. This pattern contained peaks and valleys of a stationary wave which imprinted permanently the pattern into the core of the fiber as an index of refraction modulation. Initially, the reflected light intensity is low, but after some time, it grows in intensity until almost all the light launched into the fiber is back-reflected. The growth in back-reflected light was explained in terms of a new effect called "photosensitivity".

After the inscription of the grating into de fiber's core, due the periodic modulation of the index of refraction, light guided along the core of the fiber will be weakly reflected by each grating plane by Fresnel effect. The reflected light from each grating plane will join together with the other reflections in the backward direction. This addition may be constructive or destructive, depending whether the wavelength of the incoming light meets the Bragg condition of Eq. (1).

Now, according to Eq. (1), since $\theta=90°$ and d is the distance between peaks of the interference pattern, $\lambda=2d$ for $n=1$ is the approximate wavelength of the reflection peak. That is, the fiber now acts as a dichroic mirror, reflecting part of the incoming spectrum. Equation (1), developed for vacuum, has to be adapted for silica, since the distances traveled by light are affected by the index of refraction of the fiber:

$$\lambda_B = 2n_{eff}\Lambda \tag{2}$$

Therefore the Bragg wavelength (λ_B) of an FBG is a function of the effective refractive index of the fiber (η_{eff}) and the periodicity of the grating (Λ).

The photosensitivity phenomenon in optical fibers remained unexplored for several years after its discovery, mainly due the fact that the resulted Bragg wavelength was always a function of the wavelength of the light source used and very far away from the interested region of the spectrum, namely, the third telecommunication window. However, a renewed interest appeared years later with the demonstration of the side writing technique by Gerry Meltz and Bill Morey of United Technology Research Center [3] and later on with the possibility of tuning the Bragg wavelength into the C Band of the telecom spectrum.

Equation (2), also known as the Bragg reflection wavelength, is the peak wavelength of the narrowband spectral component reflected by the FBG. The FWHM (full-width-half-maximum) or bandwidth of this reflection depends on several parameters, particularly the grating length. Typically, the FWHM is 0.05 to 0.3 nm in most sensor applications. Figure 2 shows a typical Bragg reflection peak. The lateral lobes sometimes pose problems in automatic identification of the center wavelength and in telecom applications, such as wavelength division multiplexing (WDM), these side-lobes need to be suppressed in order to reduced the separation between the optical carriers, according to ITU-T-G.694.1 (International Telecommunications Union). The side-lobes can be suppressed during the FBG fabrication by a technique known as apodization.

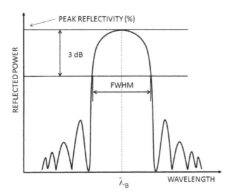

Figure 2. A typical Bragg reflection wave shape with its parameters defined.

From Eq. (2) we see that the Bragg wavelength only depends on the distance between gratings (Λ) and the effective index of refraction (n_{eff}).

Essentially, any external agent that is capable of changing Λ will displace the reflected spectrum centered at Bragg wavelength. A longitudinal deformation, due to an external force, for instance, may change both Λ and n_{eff}, the latter by the photo-elastic effect and the former by increasing the pitch of the grating. Equally, a variation in temperature can also change both parameters, via thermal dilation and thermo-optic effect respectively.

Therefore FBG is essentially a sensor of temperature and strain but, by designing the proper interface, many other measurements can be made to impose perturbation on the grating resulting in a shift in the Bragg wavelength which can then be used as a parameter transducer. Therefore, by using an FBG as a sensor we can obtain measurements of strain, temperature, pressure, vibration, displacement, etc.

Besides the influence of temperature and strain on the Bragg grating periodicity, one can also use n_{eff}, the fiber effective refractive index (RI) as a parameter transducer. The effective refractive index is an average of the RI of the core (n_{co}) and the RI of the cladding (n_{clad}) of the fiber. This parameter depends on how much the evanescent field of the core penetrates into the cladding. Since the fiber cladding diameter (125 μm) is much larger than the evanescent field, the effective RI is undisturbed by external influences. However by a corrosion of the fiber cladding by acid etching, one can reach the evanescent field which lies about 1.5 μm from the core interface. Now, the effective RI depends also on the surrounding RI, that is, the air, a gas or a liquid outside the fiber and we just created a device that can measure the RI of substances.

Since the strain or temperature measurements are encoded into wavelength shifts, these sensors are also self-calibrated because wavelength is an absolute parameter. Thus these sensors do not drift on the total light levels, losses in the connecting fibers and couplers or light source power. Additionally, the wavelength encoded nature of the output also allows the use of wavelength division multiplexing technique (WDM) by assigning each sensor to a different wavelength range of the available light source spectrum.

Using such a device and by injecting a spectrally broadband source of light into the fiber, a narrowband spectral component at the Bragg wavelength will be reflected by the grating. This spectral component will be missed in the transmitted signal, but the remainder of this light may be used to illuminate other FBGs in the same fiber, each one tuned to a different Bragg wavelength. The final result of such an arrangement is that we will have all Bragg peak reflections of each FBG back at the beginning of the fiber, each one in its specific wavelength range.

In order to calculate the sensitivity of the Bragg wavelength with temperature and strain we start from Eq. (2) and notice that the sensitivity with temperature is the partial derivative with respect of temperature:

$$\frac{\Delta\lambda_B}{\Delta T} = 2n_{eff}\frac{\partial\Lambda}{\partial T} + 2\Lambda\frac{\partial n_{eff}}{\partial T} \tag{3}$$

Substituting twice (2) in (3) we get:

$$\frac{\Delta\lambda_B}{\Delta T} = \frac{1}{\Lambda}\frac{\partial\Lambda}{\partial T}\lambda_B + \frac{1}{n_{eff}}\frac{\partial n_{eff}}{\partial T}\lambda_B$$

or rearranging,

$$\frac{\Delta\lambda_B}{\lambda_B} = \frac{1}{\Lambda}\frac{\partial\Lambda}{\partial T}\Delta T + \frac{1}{n_{eff}}\frac{\partial n_{eff}}{\partial T}\Delta T$$

The first term is the thermal expansion of silica (α) and the second term is the thermo-optic coefficient (η) representing the temperature dependence of the refractive index (dn/dT). Substituting we have:

$$\frac{\Delta\lambda_B}{\lambda_B} = (\alpha + \eta)\Delta T \tag{4}$$

The sensitivity with strain is the partial derivative of (2) with respect to displacement:

$$\frac{\Delta\lambda_B}{\Delta L} = 2n_{eff}\frac{\partial\Lambda}{\partial L} + 2\Lambda\frac{\partial n_{eff}}{\partial L} \tag{5}$$

Substituting twice (2) in (5), we have:

$$\frac{\Delta\lambda_B}{\lambda_B} = \frac{1}{\Lambda}\frac{\partial\Lambda}{\partial L}\Delta L + \frac{1}{n_{eff}}\frac{\partial n_{eff}}{\partial L}\Delta L \tag{6}$$

The first term in Eq. (6) is the strain of the grating period due to the extension of the fiber. Suppose we have a length L of a fiber with an inscribed FBG in it. If we apply a stress on the fiber of ΔL then we will have an relative strain $\Delta L/L$. At the same time if the FBG has a length L_{FBG} it will experience a strain $\Delta L_{FBG}/L_{FBG}$ but since the FBG is in the fiber, then $\Delta L_{FBG}/L_{FBG}=\Delta L/L$. Since the Bragg displacement with extension equals the displacement of the grating period with the same extension and, therefore, the first term in Eq. (6) is the unit.

The second term in Eq. (6) is the photo-elastic coefficient (ρ_e), the variation of the index of refraction with strain. In some solids, depending on the Poisson ratio of the material, this effect is negative, that is, when one expands a transparent medium, as an optical fiber for instance, the index of refraction decreases due to the decrease of density of the material. Then, when an extension is applied to the fiber, the two terms in Eq. (6) produce opposite effects, one by increasing the distance between gratings and thus augmenting the Bragg wavelength and the other by decreasing the effective RI and thus decreasing the Bragg wavelength. The combined effect of both phenomena is the classical form of the Bragg wavelength displacement with strain:

$$\frac{\Delta\lambda_B}{\lambda_B} = (1 - \rho_e)\varepsilon_z \qquad (7)$$

where ε_z is the longitudinal strain of the grating. Combining (4) and (7) together we finally end up with the sensitivity of the Bragg wavelength with temperature and strain:

$$\frac{\Delta\lambda_B}{\lambda_B} = (1 - \rho_e)\varepsilon_z + (\alpha + \eta)\Delta T \qquad (8)$$

The parameters in (8) have the following values for a silica fiber with a germanium doped core:

$$\rho_e = 0.22,$$

$$\alpha = 0.55 \times 10\text{-}6/°C,$$

and

$$\eta = 8.6 \times 10\text{-}6/°C.$$

Thus the sensitivity of the grating to temperature and strain at the wavelength range of 1550 nm, after substituting the constants in (8) are:

$$\frac{\Delta\lambda_B}{\Delta T} = 14.18 \text{ pm/°C} \qquad (9)$$

and:

$$\frac{\Delta\lambda_B}{\Delta\varepsilon} = 1.2 \text{ pm/}\mu\varepsilon \qquad (10)$$

These theoretical values, though, are not absolute as each FBG of the same fabrication batch will present slightly different sensitivities, as we will see later in the following sections.

3. Temperature compensation

Equation (8) shows that the Bragg displacement is a function of both strain and temperature. By observing only $\Delta\lambda_B$ one cannot tell if the displacement was due to strain, temperature or both. If one wants to measure only temperature, the FBG must be protected against strain which can be simply done by loosely inserting the FBG into a small-bore rigid tubing. However, if one wants to measure strain, it is very difficult to stop variation of local temperature to reach the FBG; instead, we have to compensate this variation. In order to do this we have to measure the local temperature, by a thermistor, for instance, and apply Eq. (4) to calculate the effect of temperature alone in the Bragg wavelength displacement. Then, the displacement of the Bragg wavelength due to strain alone is the total displacement observed minus the displacement due to temperature alone.

This approach is only valid if it was possible to electrically measure the temperature, which is not always the case since the local of interest could be a high voltage environment or a place with a high EMI.

The more elegant way is by the use of another FBG on the same fiber, protected against strain and at the same temperature as its neighbor. The two FBGs will be in the same fiber-optic and will provide two different Bragg reflections, one dependent on strain and temperature and the other dependent only on temperature, for compensation.

From Eq. (3) we have for the first FBG:

$$\Delta\lambda_{B1} = K_{\varepsilon 1}\Delta\varepsilon + K_{T1}\Delta T \tag{11}$$

Where

$$K_{\varepsilon 1} = (1 - \rho_e)\lambda_{B1} \tag{12}$$

$$K_{T1} = (\alpha + \eta)\lambda_{B1} \tag{13}$$

Similarly, for the other FBG we have:

$$\Delta\lambda_{B2} = K_{\varepsilon 2}\Delta\varepsilon + K_{T2}\Delta T \tag{14}$$

Where,

$$K_{\varepsilon 2} = (1 - \rho_e)\lambda_{B2} \tag{15}$$

$$K_{T2} = (\alpha + \eta)\lambda_{B2} \tag{16}$$

But since this FBG is strain free, the first term of (14) will not exist and $K_{\varepsilon 2}$ equals zero. Equations (11) and (14) can be written in matrix form:

$$\begin{bmatrix} \Delta\lambda_{B1} \\ \Delta\lambda_{B2} \end{bmatrix} = \begin{bmatrix} K_{\varepsilon 1} & K_{T1} \\ K_{\varepsilon 2} & K_{T2} \end{bmatrix} \times \begin{bmatrix} \Delta\varepsilon \\ \Delta T \end{bmatrix} \tag{17}$$

Equation (17) is called the wavelength shift matrix because its solution gives us the wavelength displacements of both FBGs as a function of temperature and strain. However, we need to find the sensing matrix that gives us the strain and temperature as a function of the wavelength displacement of each FBG. So, we multiply both sides of Eq. (17) by the inverse of the 2x2 matrix and get to:

$$\begin{bmatrix} \Delta\varepsilon \\ \Delta T \end{bmatrix} = \begin{bmatrix} K_{\varepsilon 1} & K_{T1} \\ K_{\varepsilon 2} & K_{T2} \end{bmatrix}^{-1} \times \begin{bmatrix} \Delta\lambda_{B1} \\ \Delta\lambda_{B2} \end{bmatrix} \tag{18}$$

Inverting the 2x2 matrix we have the sensing matrix:

$$\begin{bmatrix} \Delta\varepsilon \\ \Delta T \end{bmatrix} = \frac{1}{K_{\varepsilon 1}K_{T2} - K_{\varepsilon 2}K_{T1}} \begin{bmatrix} K_{T2} & -K_{T1} \\ -K_{\varepsilon 2} & K_{\varepsilon 1} \end{bmatrix} \times \begin{bmatrix} \Delta\lambda_{B1} \\ \Delta\lambda_{B2} \end{bmatrix} \tag{19}$$

In (19) we notice that if

$$K_{\varepsilon 1} K_{T2} \approx K_{\varepsilon 2} K_{T1} \qquad (20)$$

then we would not have a possible solution for Eq. (19) because equations (11) and (14) would be two almost parallel lines. This would happen, for instance, if the two FBGs had the same coefficients and Bragg wavelength reflection and would, therefore displace equally. Notice that Eq. (12) and Eq. (15), as well as Eq. (13) and Eq. (16), respectively, differ only by the Bragg wavelength. So, to avoid the redundancy in Eq. (19) we can use FBGs with Bragg reflections wide apart.

Now we can solve Eq. (19) for strain and temperature:

$$\Delta\varepsilon = \frac{1}{K_{\varepsilon 1} K_{T2} - K_{\varepsilon 2} K_{T1}} (K_{T2} \Delta\lambda_{B1} - K_{T1} \Delta\lambda_{B2}) \qquad (21)$$

$$\Delta T = \frac{1}{K_{\varepsilon 1} K_{T2} - K_{\varepsilon 2} K_{T1}} (K_{\varepsilon 1} \Delta\lambda_{B2} - K_{\varepsilon 2} \Delta\lambda_{B1}) \qquad (22)$$

Equation (21) gives the real strain of FBG 1 as measured by $\Delta\lambda_{B1}$, compensated against temperature variation measured by $\Delta\lambda_{B2}$. Equation (22) gives the temperature of the sensors. It can be used for further compensation, as for instance the thermal dilation of the metallic parts of the setup.

4. Calibration of FBG with temperature and uncertainty assessment

As it will be seen below, Eq. (9) is not an exact model for the FBG behavior under temperature variation and therefore each FBG has to be independently calibrated in order to be possible to tell the temperature by the Bragg wavelength. In this section we demonstrate the procedure to calibrate an FBG chain made of five FBGs. In this study five FBGs were submitted to temperature variations between 20°C and 85°C in order to verify and quantify the parameters of Eq. (4) [4].

In order to measure temperature, we can use as many FBG as necessary, in different Bragg wavelengths; the only precaution is that each FBG's spectrum should not overlap with its neighbor during its displacement when the temperature varies. To obtain the largest range for five FBGs we distributed them along the available range of most FBGs interrogators, that is, 1530 nm-1570 nm.

The setup used to calibrate the FBGs is shown in Figure 3. The dotted square represents the optical system comprised of a commercial Bragg Meter (Spectral Eye 400 from FOS&S) that consists of an ASE (Amplified Spontaneous Emission) broadband source used to illuminate the FBGs via Port 1 of the optical circulator. The reflection spectrum of the FBGs returns through Port 2 and is directed via Port 3 to an embedded OSA where the reflected spectrum is detected and measured. All controls and data can be accessed by a computer connected to the USB port of the interrogator.

Figure 4 shows superimposed spectra of five FBGs recorded in a temperature variation from 20°C to 85°C.

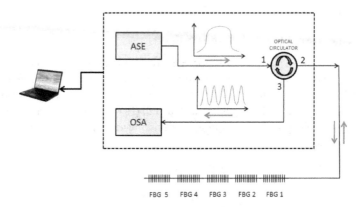

Figure 3. Schematic diagram of the measurement technique [4].

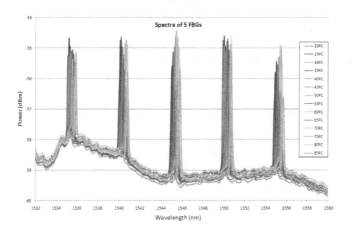

Figure 4. Superimposed spectra of five FBGs recorded in a temperature variation from 20∘C to 85∘C[4].

The procedure to calibrate the sensors followed the sequence: they were immersed simultaneously into a controlled temperature bath and the Bragg wavelengths were monitored and recorded along with the temperature given by a NIST-traceable thermometer (TD 990, Thermolink, 0.1°C resolution and ±1°C accuracy). Five sets of measurements were performed for each sensor in the range of 20°C to 85°C. Table I shows Bragg shift for each temperature and for each FBG.

From the data in Table 1 it is possible to calculate the sensitivity of each sensor, as predicted by (4) and the accuracy of the measurement chain. The graph in Figure 5 was built from the data in Table 1.

Table 2 shows a summary of the calibration parameters: the theoretical and experimental sensitivities, the correlation coefficients of the curve fittings, the root mean square errors (RMSE) and maximum residual errors.

T (°C)	Average Bragg wavelength peak (nm)				
	FBG1	FBG2	FBG3	FBG4	FBG5
25	1536,001	1540,928	1545,833	1550,819	1555,723
30	1536,067	1540,995	1545,903	1550,888	1555,793
35	1536,128	1541,059	1545,971	1550,959	1555,865
40	1536,183	1541,116	1546,03	1551,015	1555,922
45	1536,245	1541,177	1546,092	1551,083	1555,987
50	1536,310	1541,247	1546,163	1551,147	1556,055
55	1536,368	1541,308	1546,224	1551,213	1556,119
60	1536,427	1541,368	1546,284	1551,273	1556,178
65	1536,497	1541,440	1546,358	1551,345	1556,253
70	1536,557	1541,501	1546,420	1551,408	1556,317
75	1536,618	1541,566	1546,484	1551,469	1556,383
80	1536,680	1541,627	1546,549	1551,539	1556,447

Table 1. Average Bragg center wavelength of each FBG under temperature variation [4]

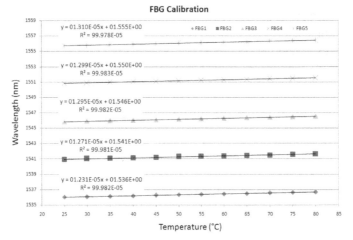

Figure 5. Wavelength shift versus temperature for each FBG [4].

FBG #	Theoretical Sensitivity (pm/°C)	Measured Sensitivity (pm/°C)	Correlation Coefficient (R^2)	RMSE (°C)	Max. Residual Error (°C)
1	14.05	12.31	0.99982	0.00311	0.003
2	14.10	12.71	0.99981	0.00328	0.005
3	14.14	12.95	0.99982	0.00327	0.005
4	14.19	12.99	0.99993	0.00320	0.006
5	14.23	13.10	0.99978	0.00363	0.007

Table 2. Calibration Parameters [4]

Figure 6 shows the error analysis for FBG1; the maximum positive error was approximately 0.004°C at 30°C. The temperature error measurements for other sensors were within the range of ±0.007°C.

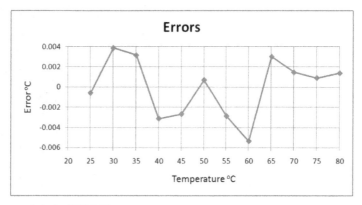

Figure 6. Error analysis for FBG 1 [4].

All correlation coefficients are very close to unity and errors much smaller than 1°C. These errors are a combination of the uncertainty of the interrogation system, (±1 pm) and of the thermometer used. Using the FBG's average sensitivity of 13 pm/°C (see Table 2), 1 pm in error means a temperature uncertainty of about 0.08°C which is much smaller that the error produced by the thermometer.

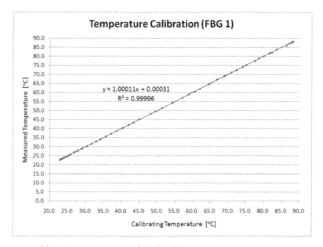

Figure 7. Temperature calibration responses of FBG 1 [4].

Notice in Table 2 that the theoretical sensitivities predicted by Equation (3) are different from those obtained in the calibration experiment. Also, Eq. (1) shows λ_B as a function of n_{eff}, the average index of refraction between the pristine fiber core and that of the Ultra-Violet (UV)-irradiated core. The FBG fabrication processes is not automatic and the radiation time for each FBG inscription is not the same as the laser is turned off by the operator when Bragg reflection appears above the desired level. The UV modifies the index of refraction of the fiber core and also modifies the values of η in each FBG differently. These results in the slightly dispersed sensitivities found above. This effect is confirmed by [5] that demonstrated a technique for changing the temperature responsivity of FBGs through increased UV exposure over the FBG.

From the data in Table 2 it is possible to calculate the relationship between wavelength and temperature for each FBG. to-one fitting accuracy to the third decimal place in temperature and a correlation coefficient $R^2=0.99996$, demonstrating a very good linearity and accuracy of FBG sensors for temperature measurements.

5. Photosensitivity in optical fibers

Photosensitivity in glass, as mentioned in Session 2, was discovered at the Communications Research Center in Canada, in 1978 by Hill and co-workers [1]. It was a new nonlinear effect in optical fibers and was called at that time of fiber photosensitivity.

A decade later after this discovery, Meltz and co-workers [3] have proposed a model for fiber photosensitivity. What motivated their model was that at first, fiber photosensitivity was detected only in fibers containing germanium as a dopant.

The model is based on the fact that when germanium-doped silica fibers are fabricated by MCVD technique, germania (GeO_2) and silica (SiO_2) in form of gases combine in high temperatures to produce the fiber. During the process though, there is a statistical probability that products like Ge-Ge, O-Ge-O, Ge_2^0 and Ge-Si might be formed, which are defects in the fiber lattice and are called in the literature as "wrong bonds". The fiber presents strong absorption peak at 245 nm, which is associated with these defects. When these defects are irradiated with UV light some absorptions bands appear and the index of refraction increases in these points.

The origins of photosensitivity and the change of the RI as a consequence have yet to be fully understood as no single model can explain the experimental results shown in the literature. So, it becomes apparent that photosensitivity is a function of several mechanisms such as photochemical, photomechanical, thermochemical, etc. [6].

One of the models that show consistencies with experimental results seems to be the compaction model based on the Lorentz-Lorenz law which states that the RI increases with material compression. This idea was pursued

These equations were fed into the software of the FBG interrogation system which returns the temperature of each sensor in a field application. Finally, it is possible to plot the calibrating temperature against measured temperature, as shown in Figure 7, presenting a one-by [7] that used UV Laser irradiation to produce thermally reversible, linear compaction in amorphous SiO_2. An accumulated, incident dose of 2000 J/cm² would produce an irreversible compaction and photoetching. The above results are in accordance with the fabrication of Type I and Type IIA FBG. Also, Laser compaction results were found to be consistent with those obtained using hydrostatic pressure. Therefore it was observed an approximately linear RI versus $\Delta V/V$ agreeing with the predictions of the Lorentz-Lorenz law.

Also, in accordance with this model of densification associated with the writing of Bragg gratings, Riant and co-workers [8] observed a transition mechanism between Type I and Type II gratings.

Notice that, when a transparent material is compressed we observe two effects that interfere with the RI. One is the increase of RI due to the increase of density of the material. The other is the photoelastic effect, which is negative for many optical media, and produces an opposite effect. However the compression produces an effect much stronger than the photoelastic effect and we normally observe an increase of RI in the irradiated parts of the FBG.

With the knowledge that the UV irradiation produces an increase on the RI we can now go further and observe physically the FBG. Figure 8 shows the core's RI along the length of the fiber (z axis) with most frequent values of the parameters. The effective RI of the core is the average index of refraction of the irradiated portion of the core and is approximately 1.45. Due to the UV radiation the variation of the RI is about $\Delta n = 10^{-4}$. The grating period (Λ) is the same as the interference pattern, about 500 nm and the FBG length L_{FBG} is around 10 mm.

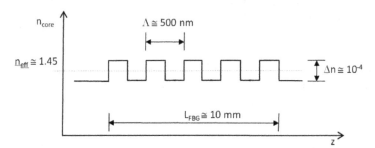

Figure 8. The Refraction Index variation of the fiber core along the length of the fiber (z axis) with most frequent values of the parameters.

The interference pattern, however, does not vary as a square wave but rather as an approximate sinusoidal waveform, which will inscribe an RI variation on the fiber of the same form, as shown in Fig. 9. In the figure, n_{eff} is the RI of the pristine fiber core, $\langle n_{eff} \rangle$ is the average RI in the FBG region, Δn_{DC} is the average amount of RI increased by the UV dose and Δn_{AC} is half of the total RI variation in the FBG.

The mathematical model of the RI in the FBG area as a function of the UV radiation dose (d) in Joules and the distance z along the fiber's axis is [9]:

$$n_{eff}(z, d) = n_{eff} + \left(\Delta n_{DC}(d) + \Delta n_{AC}(d)\sin\frac{2\pi}{\Lambda}z \right) \tag{23}$$

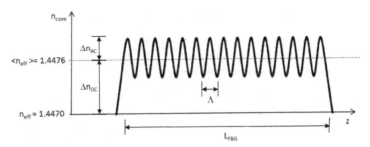

Figure 9. Variation of the Refraction Index of the fiber's core along the length of the fiber (z axis) resultant of a sinusoidal diffraction pattern [9].

Notice that both Δn_{DC} and Δn_{AC} increase with UV dose and, since the UV radiation is never zero along the diffraction region, all FBG length experiences an increase of RI.

Now we can rewrite Eq. (2) using the average RI in the FBG area:

$$\lambda_B = 2\langle n_{eff}\rangle \Lambda \tag{24}$$

For an FBG inscription using the setup of Figure 11, it is necessary first to calculate the angle φ for the desired Bragg wavelength. Then, the operator monitors the reflection spectrum until the reflectivity reaches the desired value. But since $\langle n_{eff}\rangle$ increases during the irradiation, so does λ_B, according to Eq. (24). Figure 10 shows the progression of an FBG reflection spectrum as UV dose increases.

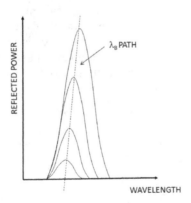

Figure 10. Progression of an FBG reflection spectrum as UV dose increases [9].

The dotted line in Figure 10 shows the path of the Bragg wavelength as UV dose increases, or same to say, the variation of $\langle n_{eff}\rangle$, according to Eq. (24).

Therefore, two relationships can be obtained from Figure 10:

$$\text{Reflectivity} = f(\text{number of shots or irradiation time}) \tag{25}$$

$$\lambda B = g(\text{number of shots or irradiation time}), \tag{26}$$

where f(*) And g(*) are arbitrary functions to be determined by curve fitting.

Both reflectivity and Bragg wavelength increase with the number of laser shots because each shot represents a certain amount of UV energy and the energy is integrated producing the stress inside the fiber core. These equations will be important when an FBG is designed.

As UV dose increases, so does reflectivity, but up to a threshold above which the reflectivity starts do lower again. This is due to the competing effects between the increase of RI due to the increase of density of the material and the photoelastic effect, which is negative for many optical media. Above the threshold limit, the photoelastic effect is stronger than the increase of density, probably because this last effect saturates while the photoelastic effect does not. The threshold is about 500 mJ/cm^2 and this value is considered to be the limit separating Type I gratings to Type IIA. While during the formation of Type I FBG, λ_B experiences a red shift (see Figure 10), during the erasure of the FBG in type IIA formation, the Bragg wavelength experiences a blue-shift. Above this limit, the FBG starts to be erased until it completely disappears.

Bragg grating in germanosilacate fibers exhibits a temperature decay dependency. Type I FBGs are found to present reasonable short term stability up to 300°C, whereas Type IIA gratings exhibit very good stability up to 500°C [6].

Therefore, after fabricating a Type I FBG it is recommended to submit them to an annealing process up to a temperature that exceeds the service temperatures of the application in order to produce an accelerated ageing.

Conventional telecommunication fibers normally present around 3.5% concentration of germania doping. These fibers will weakly respond to UV radiation ($\Delta n \cong 10^{-5}$) and will grow low reflectivity FBGs. Only fiber with 5% plus GeO concentration present photosensitivity enough to be useful for FBG fabrication, but are more expensive than conventional telecommunication fibers. Those fibers with up to 30% of dopant concentration are produced by several fiber makers such as Nufern, Fibercore and IPTH.

Another dopant used is GeO co-doped with boron; these fibers present an enhanced sensitivity however causing an increase in losses. Therefore, boron co-doped fibers are not good for long sensing distance, they are limited to some few meters only in contrast with pure GeO doped fibers that can be used to remotely monitor parameters which are several kilometers away from the interrogation system.

Another way of enhancing the fiber sensitivity is by hydrogen diffusion into the fiber core. The mechanism causing an increased sensitivity is thought to be due the reaction of H_2 with GeO. In highly doped fibers there is a significant concentration of Ge-O-Ge bonds. H_2 reacts with these bonds resulting in the formation of Ge-OH, which absorbs UV radiation and therefore increasing the internal stress into the core of the fiber [10].

The hydrogen diffusion is accomplished by leaving the fiber into a tight enclosure with hydrogen at high pressure. Pressures from 20 atm to 750 atm can be used but most commonly 150 atm. Apart from increasing the fiber photosensitivity, hydrogen loading allows the fabrication of FBG in any germane silicate and germanium free fibers.

When a hydrogen loaded fiber is taken out of the high pressure vessel it is as soft as cotton string. However, by heating the fiber after the exposition, the hydrogen diffuses out in a few minutes.

Hydrogen loading can be also accomplished by a technique known as flame brushing. This technique consists in burning for 20 minutes the fiber by a hydrogen-oxygen flame reaching temperatures of 1700ºC. At high temperature, the excess of hydrogen in the mixture diffuses into the fiber. The advantage of frame brushing is that it is possible to sensitize conventional telecommunications fibers. The disadvantage, however, is that the flame burns the fiber acrylate buffer in an area larger than that of the FBG itself which demands a posterior fiber recoating.

6. Fabrications techniques

As mentioned above the interest in FBG started with the possibility of inscribing the grating sideways as demonstrated by Meltz [3] and with the possibility of tuning to a desired wavelength along the telecom band. From then on, many FBG applications appeared first in telecom such as add/drop, dense wavelength division multiplexing (DWDM) mux/demux, filters, lasers, and so on. Later, with the telecommunications devices and equipments decreasing prices, FGBs started to be used as sensors in a commercial basis.

The first technique used to inscribe a FBG in the fiber was the interferometer and it is used in many different configurations (Figure 11 shows a basic interferometer). A laser beam is divided in two by a beam splitter or a prism. Each part is reflected in mirrors to meet again to form a interference pattern over the fiber to be inscribed. Cylindrical lenses concentrate the beams in the inscribing area of the fiber, about 5 mm by 200 μm in order to increase the density of the UV dose.

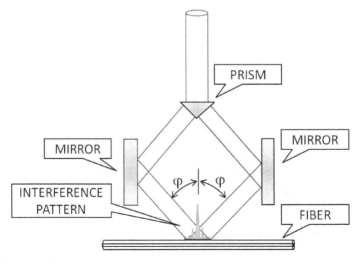

Figure 11. A basic interferometer.

The period of the interference pattern (Λ) depends on the wavelength of the light used for writing (λ_{Laser}) and also on the half-angle between the two interfering beams (φ) as shown in Eq. (27).

$$\Lambda = \frac{\lambda_{Laser}}{2 \sin\varphi} \tag{27}$$

There are some disadvantages when using such arrangement as, for instance, the difficult to align the beams, the necessity to achieve a very good spatial coherence on the laser and problems associated with air flow which slightly modifies the RI of the air distorting the wave front of the beams. This effect could lead to an FBG of poor quality. The advantage is that one can adjust the mirrors in order to vary the grating period to virtually any value around the telecommunication band.

Nowadays we rely on the phase-mask technique which is a diffractive optical element that spatially modulates the UV beam with period Λ_{pm}. The phase masks are formed in a fused silica substrate by a holographic technique or electron beam lithography.

When a laser beam is incident to the phase mask a diffraction occurs and the beam is divided into several diffraction orders. The zero order is suppressed to less than 3%, but the +1 and -1 orders prevail with most of the remaining power. These two orders start from the same point on the other side of the phase mask but are divergent. At the near field an interference pattern is produced as the two orders cross each other, with a period

$$\Lambda = \frac{\Lambda_{pm}}{2} \tag{28}$$

The optical fiber is placed in contact or in close proximity to the phase mask, inside the near field where the interference pattern is produced as shown in Figure 12. An increased power density can be achieved by the use of a cylindrical lens parallel to the fiber, before the phase mask.

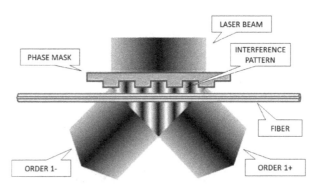

Figure 12. A laser beam inscribing gratings in a optical fiber through a phase mask.

The advantage of the phase mask is that its setup is much simpler because there is no need for the laser to have a good coherence and there are no mirrors to align. However, as the phase mask is such a fragile optical element, the close proximity of the fiber to the phase mask surface can scratch it. If the distance between the phase mask is increased by a few millimeters the fiber will be illuminated by a narrower interference and the FBG will be accordingly narrower. Another disadvantage of the phase mask technique is that the periodicity of the FBG inscribed is fixed by the one of the phase masks, according to Equation (28).

An alternative setup is shown in Figure 13 in which the phase mask is far away from the fiber and the two mirrors redirect the +1 and -1 refracted orders back to the fiber.

The advantage of such setup is that one can adjust the Bragg wavelength by the angles of the mirrors.

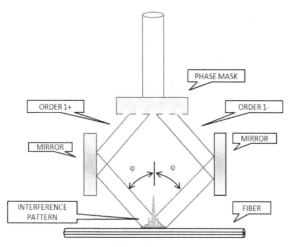

Figure 13. Most common setup used with a phase mask.

7. FBG fabrication parameters

When specifying an FBG the following parameters must be known:

- Central Brag wavelength, λ_B
- FBG width, FWHM
- Reflectivity

Bragg wavelength(λ_B) depends essentially on the phase mask periodicity or on the laser wavelength, and on the intersection half-angle in the case when an interferometer setup is used (Figure 11). However, the UV dose also modifies the Bragg wavelength, according to Eq. (25).

FWHM depends on the FBG length and on the UV dose, according to the following Equation [6]:

$$\text{FWHM} = \lambda_B s \sqrt{\left(\frac{\Delta n}{2n_{\text{eff}}}\right)^2 + \frac{1}{N}^2} \qquad (29)$$

where, Δn is the amplitude of the induced RI in the fiber, $\Delta n = 2 \times \Delta n_{AC}$ (see Figure 9), λ_B is the final Bragg wavelength, s is 1 for strong reflection grating with reflectivity close to 100%, or 0.5 for weak gratings, N is the number of grating planes, $N = L_{FBG}/\Lambda$.

Reflectivity, as it has been seen in later sessions, is a function of the UV dose in J/m^2, the amount of Germania doping in the fiber and the hydrogenation processes. Equation (26) can be useful for predicting the reflectivity and this parameter can also be adjusted by varying the FBG length. This is accomplished by adjusting the laser beam width. One can produce lengths as small as 2.5 mm obtaining, thus very low reflectivity and lengths as large as 15 mm obtaining a reflectivity around 100%.

Therefore, when projecting an FBG inscription, one has to carry out an inverse engineering to preview how much λ_B will displace during the inscription to the desired reflectivity and decrease this value to the desired Bragg wavelength in order to adjust the mirrors accordingly.

As none of the above parameters are not precisely known, the best way to know the exact time of irradiation is by experimental tests.

8. Interrogation techniques for FBG sensors

The main challenge when it comes to FBG sensing is the method to demodulate its wavelength changes. The use of FBG sensors is connected to the development of techniques to interrogate these sensors and detect Bragg wavelengths shifts as a function of the parameter being measured.

The easiest way to interrogate an FBG is by the use of an optical spectrum analyzer (OSA) which performs a direct measurement of the reflection spectrum of the FBG. The other method is based on the conversion of wavelength variations into optical power intensities.

The technique using an optical spectrometer is very simple. The interrogation system consists of a broadband optical source which illuminates the FBGs. Their reflected peaks, which are represented by each Bragg wavelength λ_B, are directed to the OSA and monitored by a computer, as shown in Figure 3.

Although being a simple demodulation technique, it presents some disadvantages: first, the commercial OSAs are heavy and expensive equipment, besides being inappropriate for field applications. Moreover, most of the spectrum analyzers are limited to static measurements, so they don't have sufficient resolution concerning the response time when a number sensors are being interrogated. A conventional OSA will present an accuracy of about ±50 pm, which would produce errors of about ±4°C and ±60 με, according to equations (9) and (10). For most applications these errors are unacceptable if compared to conventional resistive temperature detectors (RTD) and strain gauges. To detect small variations in wavelength the development of new techniques must also ensure essential characteristics such as static and dynamic measurements, real-time measurements, accuracy, resolution, and low cost, all necessary conditions for field applications. In conclusion this technique is only useful for laboratory applications and tests.

The setup shown in Figure 3 is also commercially available by a few companies as standalone equipment and appropriated to go to the field. In this case they feature resolutions as good as ±2 pm and can be programmed to monitor specific parameters as pressure, strain, temperature, etc. However, due to the high cost, they will be applicable to solve monitoring needs of industries only if the project included as many sensors as possible to be monitored by one single unit to have the total price divided by the number of sensors.

The other demodulation technique uses a Fabry-Perot (FP) tunable filter. Although the interferometric FP filter method is a consolidated technology, showing high resolution and accuracy, it still presents a moderate cost. The tunable optical filter scheme is based on a Fabry-Perot extrinsic cavity, which is adjusted by mirrors and by varying the internal cavity of a PZT crystal by means of an external power supply, enabling the filter adjustment and selection of the desired wavelength. This relationship between the changes in the filter wavelength as a function of an applied voltage is linear. Defects in the geometry of the lens during the filter manufacturing process can cause instability in the measurement system, so that the optical spectrum of the filter is not entirely symmetrical.

The demodulation setup using a FP filter is shown in Figure 14. A broadband light source was used to illuminate the FBG sensor via an optical circulator. The reflected spectrum of the sensor passes through the FP tunable filter with a 0.89 nm bandwidth.

This demodulation technique is based on the same principle of an FM radio signal demodulated by an edge filter. The signal waveband is made to vary at the wedge of a filter that will transmit a variable power proportional to the variation of the signal frequency. In our case, proportional to Bragg shift. In reality, the transmitted signal through the filter is proportional to the convolution between the signal power and the filter response.

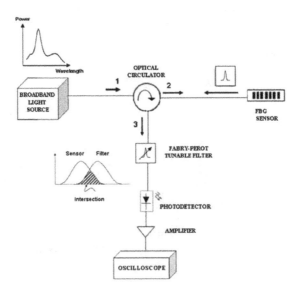

Figure 14. The interrogation setup using a Fabry-Perot filter (adapted from [11]).

The optimum position of the center wavelength of the FP filter is chosen by an algorithm described by [11]. The dashed area on the spectra drawing (inset in Figure 14) is the intersection between the spectrum of the reflected signal and the band pass of the FP filter. The integral of this area represents the total light power that reaches the photodetector.

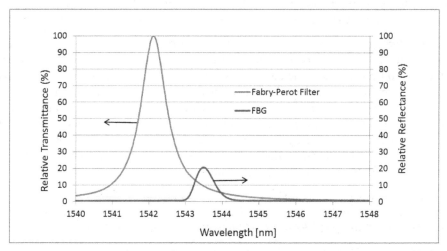

Figure 15. Spectral curves for the Fabry-Perot filter, $F_{FP}(\lambda)$, and the FBG, $F_{FBG}(\lambda)$.

The spectral curves for the FP filter, $FFP(\lambda)$, and for the FBG, $F_{FBG}(\lambda)$, are shown in Figure 15, where the sensor is at quiescent state. The vertical axes show the relative transmittance of the FP filter and the relative reflectance of the FBG sensor, respectively. The numerical convolution $F_{FP}(\lambda)*F_{FBG}(\lambda)$ represents the available power to the photodetector as a function of the wavelength shift. The convolution curve is shown in Figure 16.

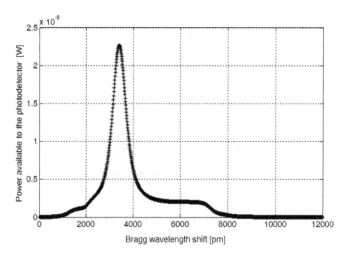

Figure 16. The convolution between $F_{FP}(\lambda)*F_{FBG}(\lambda)$.

Instead of an FB filter, it is possible to use another FBG, in this case used as a dichroic mirror, differently from the FP filter which acts by light transmission. A broadband light source injects light into port 1 of the optical circulator 1. The light circulates to port 2, illuminating the FBG sensor. The reflection spectrum of the FBG sensor is deviated to port 3, and enters through port 1 of circulator 2. Circulator 2 deviates the signal to the twin FBG filter, through port 2. Only the superimposed wavelengths (inset graphic) reflect back to circulator 2, which deviates the light to the photodetector through port 3.

This demodulation scheme is very simple, and reduces the cost of the setup implementation; however, the twin FBG must be manufactured at an exact wavelength to provide an optimized operation procedure.

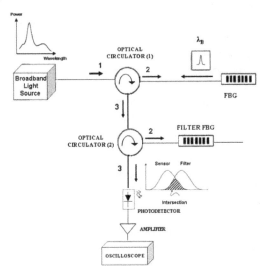

Figure 17. Schematic diagram of experiment for AC voltage measurement by using the twin grating filter technique [11].

Author details

Marcelo M. Werneck, Regina C. S. B. Allil, Bessie A. Ribeiro and Fábio V. B. de Nazaré

Instrumentation and Photonics Laboratory, Electrical Engineering Program, Universidade Federal do Rio de Janeiro (UFRJ), RJ, Brazil

Regina C. S. B. Allil

Division of Chemical, Biological and Nuclear Defense, Biological Defense Laboratory, Brazilian Army Technological Center (CTEx) RJ, Brazil

9. References

[1] Hill, K.O.; Fujii, Y.; Johnson, D. C.; Kawasaki, B. S., "Photosensitivity in optical fiber waveguides: application to reflection fiber fabrication", Appl. Phys. Lett. 32 (10): 647, 1978.

[2]   Hill, K.O., "Photosensitivity in Optical Fiber Waveguides: From Discovery to Commercialization" IEEE Journal on Selected Topics in Quantum Electronics, VOL. 6, NO. 6, pp. 1186-1189, November/December 2000

[3]   Meltz, G., Morey, W. W. and Glenn, W. H., "Formation of Bragg gratings in optical fibers by a transverse holographic method", Opt. Lett. 14 (15): 823, 1989.

[4]   Werneck, M. M., Allil, R. C. and Ribeiro, B. A., "Calibration and Operation of a Fiber Bragg Grating Temperature Sensing System in a Grid-Connected Hydrogenerator", IET Science, Measurement & Technology, accepted to publication on September, 2012.

[5]   Gwandu, B. A. L. and W. Zhang, W., "Tailoring the temperature responsivity of fibre Bragg gratings", Proceedings of IEEESensors, DOI: 10.1109/ICSENS.2004.1426454, pp. 1430-1433, Volume 3, 2004.

[6]   Othonos, A., Kalli, K., "Fiber Bragg Gratings – Fundamentals and Applications in Telecommunications and Sensing", Artech House, 1999.

[7]   Fioria, C., and Devinea, R. A. B., "Ultraviolet Irradiation Induced Compaction and Photoetching in Amorphous Thermal SiO2". MRS Proceedings of the Fall Meeting, Volume 61, 1985.

[8]   Riant, I., Borne, S., Sansonetti, P., Poumellec, B. "Evidence of densification in UV-written Bragg gratings in fibres", in Photosensitivity and Quadratic Nonlinearity in Glass Waveguides: Fundamentals and Applications, pp: 52-55, Postconference Edition, Optical Society of America, 1995.

[9]   Jülich, F. and Roths, J., "Determination of the Effective Refractive Index of Various Single Mode Fibres for Fibre Bragg", Proceedings of SENSOR+TEST Conference (OPTO-2009), Nürnberg, pp 119-124, 2009.

[10] Lemaire, P. J., Hill, M. and Erdogan, T., "Hydrogen-enhanced UV photosensitivity of optical fibers: Mechanisms and reliability", in Photosensitivity and Quadratic Nonlinearity in Glass Waveguides Fundamentals and Applications, September 9-11, 1995, Portland, Oregon, Technical Digest Series, Volume 22, pp 78-81.

[11] Ribeiro, B. A., Werneck, M. M. and Silva-Neto, J. L., "A Novel Optimization Algorithm to Demolutate a PZT-FBG sensor in AC High Voltage Measurements", in review at IEEE Sensors Journal, September, 2012.

Chapter 2

Fabrication and Mode Coupling of Long-Period Fiber Grating by Winding a Wire Around an Optical Fiber Fixed to a Cylindrical Metal Rod

Jonghun Lee, Cherl-Hee Lee, Kwang Taek Kim and Jaehee Park

Additional information is available at the end of the chapter

1. Introduction

A long-period fiber grating (LPFG) couple light from the fundamental guided core mode to co-propagating cladding modes at specific resonance wavelengths. A LPFG can be fabricated by induction of periodic modulations of the refractive index along the core of a single-mode optical fiber. The pitch of a typical LPFG ranges from 100 μm to 1000 μm, which is larger than that of a fiber Bragg gratings (FBGs) by more than two orders of magnitude. The transmission spectrum of a LPFG consists of a number of rejection bands at the resonance wavelengths, like as a band-rejection filter. LPFGs were first proposed and demonstrated by Vengsarkar and others [1] as band-rejection filters, due to their high sensitivity as well as low insertion loss and return loss, LPFGs are becoming more and more popular as simple and versatile optical components for mode converters [2], gain flattening filters [3], and fiber-optic sensors [4]. Many methods have been demonstrated for the fabrication of LPFGs. An ultra-violet (UV) writing method, which causes the index change of a photosensitive Ge-doped silica fiber core by UV light radiation, was successfully applied to the writing of LPFGs [5]. The amount of index change depends on the intensity and the duration of the UV laser. Because of the symmetrically-formed index modulation in the core, only axially symmetric cladding modes are coupled in a UV-written LPFG. Many non-UV methods have also been demonstrated for the fabrication of LPFGs, such as the electric-discharge writing method [6], divided coil heaters method [7], microbending method [8-9], and mechanically induced method [10-11-12]. However, all of these methods involve expensive fabrication equipments or additional fixed devices, thereby increasing the installation costs. Thus, the inexpensive manufacture of small LPFGs has been major barrier for the realization of practical optical communication systems. Recently, winding wires has been presented as a new fabrication method to form LPFGs on an optical fiber [13-14]. In

[13,14], since the coupling wavelength of the grating is defined by the period of the v-grooved tube, this requires very difficult and delicate fabrication operations. Thus, a new simple and flexible fabrication process was presented for the fabrication of low-cost LPFGs that does not require any formation of periodic grooves [15]. The wire-winding long-period fiber grating (WW-LPFG) presented in [15] is based on the periodic winding of a wire around an optical fiber attached to a cylindrical metal rod without any grooves. The periodic pressure from the winding wire on an optical fiber creates a periodic change in the refractive index of the optical fiber core, and makes mechanically-induced LPFGs on the optical fiber. A fabricated WW-LPFG with high tension of the winding wire on the optical fiber induces a high index variation on the fiber core. Thus, the coupling coefficient of the WW-LPFG was calculated according to the refractive index variation of the fiber core. The new approach is presented for the easy fabrication of wire-winding long-period gratings (WW-LPFGs) based on the periodic winding a wire around an optical fiber attached to a cylindrical rod. The periodic pressure of the wire induces change of the refractive index of the optical fiber core; thus, resonance wavelengths of the grating can be easily controlled according to the winding-wire pitch controlled by a microprocessor. At the winding-wire pitch of 500 μm, the spectra show resonance-wavelength dips corresponding to the LP_{02}, LP_{03}, LP_{04}, and LP_{05} cladding modes; the resonance wavelengths go to longer wavelengths at the increased the winding pitch of 510 μm. These are in good agreement with theoretical values. The polarization dependence of a resonance wavelength shift of 9 nm and transmission power difference of 2.5 dB was shown at the 520 μm grating pitch. When the diameter of the cylindrical metal rod was smaller, or higher tension from the winding pressure was applied, stronger mode coupling resulted in deeper dips in the transmission spectra according to the induced higher index variation.

2. Properties of long-period fiber grating

2.1. Effective index

The electric field in a cylindrical coordinate system has three independent vector components as a function of r, φ, and z:

$$E(r,\phi,z) = \hat{r}E_r(r,\phi,z) + \hat{\phi}E_\phi(r,\phi,z) + \hat{z}E_z(r,\phi,z). \tag{1}$$

Unlike the E_r and E_ϕ component, the E_z component does not couple to the other components of E_r and E_ϕ; therefore, the scalar wave equation for E_z can exist independently and easily be solved. Thus, the other fields, E_r and E_ϕ, are derived easily through the calculated E_z component. The wave equation for the z component of the field vectors is given as follows in the homogeneous region:

$$\left(\nabla^2 + k^2\right)\begin{Bmatrix} E_z \\ H_z \end{Bmatrix} = 0, \tag{2}$$

where $k^2 = \omega^2 \mu \varepsilon$ and ∇^2 is the Laplacian operator in the cylindrical coordinate given by

Fabrication and Mode Coupling of Long-Period Fiber Grating by Winding a Wire Around an Optical
Fiber Fixed to a Cylindrical Metal Rod

27

$$\nabla^2 = \frac{\partial^2}{\partial r^2} + \frac{1}{r}\frac{\partial}{\partial r} + \frac{1}{r^2}\frac{\partial^2}{\partial \phi^2} + \frac{\partial^2}{\partial z^2}, \tag{3}$$

Every component of the fields, including E_z, is assumed to have the angular frequency ω and the wavenumber of β to the longitudinal direction z-axis, and Eqn. (2) can be written as

$$\frac{\partial^2 E_z}{\partial r^2} + \frac{1}{r}\frac{\partial E_z}{\partial r} + \frac{1}{r^2}\frac{\partial^2 E_z}{\partial \phi^2} + \left[k^2 n(r,\phi)^2 - \beta^2\right]E_z = 0$$
$$\frac{\partial^2 H_z}{\partial r^2} + \frac{1}{r}\frac{\partial H_z}{\partial r} + \frac{1}{r^2}\frac{\partial^2 H_z}{\partial \phi^2} + \left[k^2 n(r,\phi)^2 - \beta^2\right]H_z = 0, \tag{4}$$

where $n(r,\phi)$ is the refractive index of an optical fiber at the transverse axis. Here, E_z and H_z can be expressed with the form of the Bessel differential function of order l, $l = 0,1,2,3,...$, so E_z and H_z are single-valued function of ϕ:

$$E_z = \begin{cases} AJ_l(\kappa r)e^{jl\phi}e^{-j\beta z}, (0 \le r \le a) \\ CK_l(\gamma r)e^{jl\phi}e^{-j\beta z}, (r > a) \end{cases}$$
$$H_z = \begin{cases} BJ_l(\kappa r)e^{jl\phi}e^{-j\beta z}, (0 \le r \le a) \\ DK_l(\gamma r)e^{jl\phi}e^{-j\beta z}, (r > a) \end{cases} \tag{5}$$

At Maxwell's curl equations, Eqn. (6), the transverse electromagnetic fields are related to E_z and H_z:

$$\nabla \times E = -j\omega\mu H, \nabla \times H = j\omega\varepsilon E, \tag{6}$$

Substituing Eqn. (5) to Eqn. (6), all the transverse components are obtained as follows:

At the core region (r<a):

$$E_r = -\frac{j\beta}{\kappa^2}\left[A\kappa J_l'(\kappa r) + B\frac{j\omega\mu l}{\beta r}J_l(\kappa r)\right]e^{jl\phi}e^{-j\beta z}$$
$$E_\phi = -\frac{j\beta}{\kappa^2}\left[A\frac{jl}{r}J_l(\kappa r) - B\frac{\omega\mu}{\beta}\kappa J_l'(\kappa r)\right]e^{jl\phi}e^{-j\beta z}$$
$$H_r = -\frac{j\beta}{\kappa^2}\left[-A\frac{j\omega\varepsilon_{co}l}{\beta r}J_l(\kappa r) + B\kappa J_l'(\kappa r)\right]e^{jl\phi}e^{-j\beta z}$$
$$H_\phi = -\frac{j\beta}{\kappa^2}\left[A\frac{\omega\varepsilon_{co}}{\beta}\kappa J_l'(\kappa r) + B\frac{jl}{r}J_l(\kappa r)\right]e^{jl\phi}e^{-j\beta z} \tag{7}$$

At the cladding region (r>a):

$$E_r = \frac{j\beta}{\gamma^2}\left[C\gamma K_l'(\gamma r) + D\frac{j\omega\mu l}{\beta r}K_j(\gamma r)\right]e^{jl\phi}e^{-j\beta z}$$

$$E_\phi = \frac{j\beta}{\gamma^2}\left[C\frac{jl}{r}K_l(\gamma r) - D\frac{\omega\mu}{\beta}\gamma K_l'(\gamma r)\right]e^{jl\phi}e^{-j\beta z}$$

$$H_r = \frac{j\beta}{\gamma^2}\left[-C\frac{j\omega\varepsilon_{cl}l}{\beta r}K_l(\gamma r) + D\gamma K_l'(\gamma r)\right]e^{jl\phi}e^{-j\beta z}$$

$$H_\phi = \frac{j\beta}{\gamma^2}\left[C\frac{\omega\varepsilon_{cl}}{\beta}\gamma K_l'(\gamma r) + D\frac{jl}{r}K_l(\gamma r)\right]e^{jl\phi}e^{-j\beta z}$$

(8)

where $\varepsilon_{co}=\varepsilon_0 n_{co}^2$, $\varepsilon_{cl}=\varepsilon_0 n_{cl}^2$, and

$$J_l'(\kappa r) = \frac{dJ_l(\kappa r)}{d(\kappa r)}, K_l'(\gamma r) = \frac{dK_l(\gamma r)}{d(\gamma r)}$$

(9)

The boundary condition that E_z and H_z should be continuous at r=a leads to

$$C = A\frac{J_l(\kappa a)}{K_l(\gamma a)}, D = B\frac{J_l(\kappa a)}{K_l(\gamma a)}$$

(10)

The boundary condition that E_ϕ should be continuous at r=a leads to

$$A\frac{jl}{a}\left(\frac{1}{\kappa^2} + \frac{1}{\gamma^2}\right) = B\frac{\omega\mu}{\beta}\left[\frac{1}{\kappa}\frac{J_l'(\kappa a)}{J_l(\kappa a)} + \frac{1}{\gamma}\frac{K_l'(\gamma a)}{K_l(\gamma a)}\right]$$

(11)

The boundary condition that H_ϕ should be continuous at r=a leads to

$$A\left[\frac{\omega\varepsilon_{co}}{\beta}\frac{1}{\kappa}\frac{J_l'(\kappa a)}{J_l(\kappa a)} + \frac{\omega\varepsilon_{cl}}{\beta}\frac{1}{\gamma}\frac{K_l'(\gamma a)}{K_l(\gamma a)}\right] = -B\frac{jl}{a}\left(\frac{1}{\kappa^2} + \frac{1}{\gamma^2}\right)$$

(12)

Equation (11) and (12) yield the following equation determining the effective index and propagation constant:

$$\frac{\beta^2 l^2}{a^2}\left(\frac{1}{\kappa^2} + \frac{1}{\gamma^2}\right)^2 = \left(\frac{1}{\kappa}\frac{J_l'(\kappa a)}{J_l(\kappa a)} + \frac{1}{\gamma}\frac{K_l'(\gamma a)}{K_l(\gamma a)}\right)\left(k_0^2 n_{co}^2\frac{1}{\kappa}\frac{J_l'(\kappa a)}{J_l(\kappa a)} + k_0^2 n_{cl}^2\frac{1}{\gamma}\frac{K_l'(\gamma a)}{K_l(\gamma a)}\right)$$

(13)

If weakly guiding condition, $n_{co}\cong n_{cl}$, Eqn. (13) is simplified to

$$n_{eff}^2\frac{l^2}{a^2}\left(\frac{1}{\kappa^2} + \frac{1}{\gamma^2}\right)^2 = \left(\frac{1}{\kappa}\frac{J_l'(\kappa a)}{J_l(\kappa a)} + \frac{1}{\gamma}\frac{K_l'(\gamma a)}{K_l(\gamma a)}\right)^2$$

(14)

By using the Bessel function identity,

$$\frac{J_l'(\kappa a)}{\kappa J_l(\kappa a)} = \pm \frac{J_{l\mp 1}(\kappa a)}{\kappa J_l(\kappa a)} \mp \frac{l}{\kappa^2}, \frac{K_l'(\gamma a)}{\gamma K_l(\gamma a)} = \mp \frac{K_{l\mp 1}(\gamma a)}{\gamma K_l(\gamma a)} \mp \frac{l}{\gamma^2},$$

(15)

we can get

$$\left(\pm \frac{J_{l\mp 1}(\kappa a)}{\kappa a J_l(\kappa a)} \mp \frac{l}{(ka)^2} \right) + \left(\mp \frac{K_{l\mp 1}}{\gamma a K_l} \mp \frac{l}{(\gamma a)^2} \right) = \pm l \left[\frac{1}{(\kappa a)^2} + \frac{1}{(\gamma a)^2} \right]$$

(16)

From Eqn. (16) two different equations corresponding to the two values of +*l*,-*l* are obtained and the eigenvalues resulting from these two equations yield the two classes of solutions designated as the EH modes and HE modes. The characteristic equations of the EH and HE modes correspond to the upper sign and the lower sign, respectively.

For EH mode:

$$\frac{1}{\kappa a} \frac{J_{l+1}(\kappa a)}{J_l(\kappa a)} = -\frac{1}{\gamma a} \frac{K_{l+1}(\gamma a)}{K_l(\gamma a)}$$

(17)

For HE mode:

$$\frac{1}{\kappa a} \frac{J_0(\kappa a)}{J_1(\kappa a)} = \frac{1}{\gamma a} \frac{K_0(\gamma a)}{K_1(\gamma a)}, (l=1)$$

$$\frac{1}{\kappa a} \frac{J_{l-1}(\kappa a)}{J_{l-2}(\kappa a)} = -\frac{1}{\gamma a} \frac{K_{l-1}(\gamma a)}{K_{l-2}(\gamma a)}, (l \geq 2)$$

(18)

Some characteristic equations among the TE, TM, EH, and HE modes are the same under the weakly guiding approximation, and the degenerate modes are classified as approximated linear polized (LP*lm*) modes:

$$\frac{1}{\kappa a} \frac{J_l(\kappa a)}{J_{l-1}(\kappa a)} = -\frac{1}{\gamma a} \frac{K_l(\gamma a)}{K_{l-1}(\gamma a)}$$

(19)

Figure 1 shows a graphical determination of the effective index of LP01 core mode and Fig. 2 shows the mode intensity of LP01 core mode. The cladding modes can be obtained by applying the boundary continuity conditions at the core-cladding boundary and the cladding-air boundary. Erdogan [16] proposed an accurate description of the mode propagation in the cladding, in which the cladding modes are not approximated as being linearly polarized. Applying the core-cladding boundary condition, the characteristic equation for a cladding mode was given by

$$\zeta_0 = \zeta_0'$$

(20)

$$\zeta_0 = \frac{1}{\sigma_2} \frac{u_2\left(JK + \frac{\sigma_1\sigma_2 u_{21}u_{32}}{n_2^2 a_1 a_2}\right)p_l(a_2) - Kq_l(a_2) + Jr_l(a_2) - \frac{1}{u_2}s_l(a_2)}{-u_2\left(\frac{u_{32}}{n_2^2 a_2}J - \frac{u_{21}}{n_1^2 a_1}K\right)p_l(a_2) + \frac{u_{32}}{n_1^2 a_2}q_l(a_2) + \frac{u_{21}}{n_1^2 a_1}r_l(a_2)} \tag{21}$$

$$\zeta_0' = \sigma_1 \frac{u_2\left(\frac{u_{32}}{a_2}J - \frac{n_3^2 u_{21}}{n_2^2 a_1}K\right)p_l(a_2) + \frac{u_{32}}{a_2}q_l(a_2) + \frac{u_{21}}{a_1}r_l(a_2)}{u_2\left(\frac{n_3^2}{n_2^2}JK + \frac{\sigma_1\sigma_2 u_{21}u_{32}}{n_1^2 a_1 a_2}\right)p_l(a_2) - \frac{n_3^2}{n_1^2}Kq_l(a_2) + Jr_l(a_2) - \frac{n_2^2}{n_1^2 u_2}s_l(a_2)} \tag{22}$$

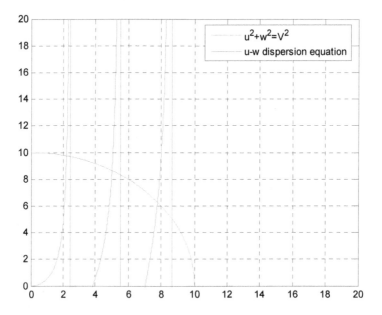

Figure 1. Graphical determination of the effective index of core mode.

Figure 3 shows the graphical determination of the effective indice of cladding modes according to Eqn. (20), which are determined by the crossing points of the left-hand side equation and the right-hand side equation.

Fabrication and Mode Coupling of Long-Period Fiber Grating by Winding a Wire Around an Optical Fiber Fixed to a Cylindrical Metal Rod

31

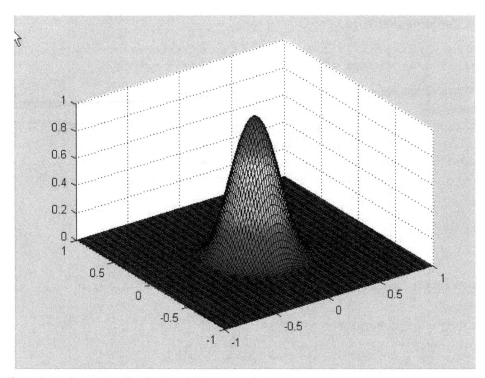

Figure 2. Mode intensity distribution of LP₀₁ core mode

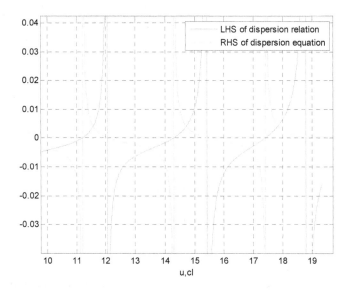

Figure 3. Graphical determination of the effective index of cladding modes.

2.2. Coupling coefficient

The transverse electric fields are described by a linear superposition of forward-propagating and backward-propagating guided modes of the unperturbed optical fiber:

$$E(x,y,z) = \left[A^+(z)e^{-j\beta z} + A^-(z)e^{j\beta z} \right] e_t(x,y), \tag{23}$$

where $A^+(z)$ and $A^-(z)$ are slowly varying amplitudes traveling in the +z and –z directions, respectively; β is the propagation constant; and $e_t(x,y)$ is the transverse mode field. The mode coupling of the electric field occurs at the gratings of the optical fiber core, which acts as a perturbation in the refractive index. Thus, the coupled mode equation can be written as

$$
\begin{aligned}
\frac{dA^+}{dz} &= -jA^+ K e^{-j\left(\beta^+ - \beta^+\right)z} - jA^- K e^{j\left(\beta^+ + \beta^-\right)z} \\
\frac{dA^-}{dz} &= jA^+ K e^{-j\left(\beta^+ + \beta^-\right)z} + jA^- K e^{j\left(\beta^- - \beta^-\right)z}
\end{aligned}
\tag{24}
$$

where $K(z)$, the general transverse coupling coefficient between modes, is given

$$
\begin{aligned}
K(z) &= \frac{\omega}{4} \int_{-\infty}^{\infty}\int_{-\infty}^{\infty} \Delta\varepsilon(x,y,z) e_t(x,y) \cdot e_t^*(x,y) dxdy \\
&= \frac{\omega}{4} \int_{-\infty}^{\infty}\int_{-\infty}^{\infty} \left\{ 2n_{co}\delta n_{co}\left[1 + v\cos\left(\frac{2\pi}{\Lambda}z + \phi\right)\right]\right\} e_t(x,y) \cdot e_t^*(x,y) dxdy \\
&= \frac{\omega n_{co}}{2} \int_{-\infty}^{\infty}\int_{-\infty}^{\infty} \delta n_{co} e_t(x,y) \cdot e_t^*(x,y) dxdy + \frac{\omega n_{co}}{2}\frac{v}{2}\int_{-\infty}^{\infty}\int_{-\infty}^{\infty} 2\cos\left(\frac{2\pi}{\Lambda}z + \phi\right) e_t(x,y) \cdot e_t^*(x,y) dxdy \\
&= \frac{\omega n_{co}}{2} \int_{-\infty}^{\infty}\int_{-\infty}^{\infty} \delta n_{co} e_t(x,y) \cdot e_t^*(x,y) dxdy + \frac{v}{2}\left[\frac{\omega n_{co}}{2}\int_{-\infty}^{\infty}\int_{-\infty}^{\infty} e_t(x,y) \cdot e_t^*(x,y) dxdy\right]\left[2\cos\left(\frac{2\pi}{\Lambda}z + \phi\right)\right] \\
&= \sigma(z) + \left[\frac{v}{2}\sigma(z)\right]\left[2\cos\left(\frac{2\pi}{\Lambda}z + \phi\right)\right] \\
&= \sigma(z) + \kappa\left[2\cos\left(\frac{2\pi}{\Lambda}z + \phi\right)\right]
\end{aligned}
\tag{25}
$$

where $\Delta\varepsilon$ is the permittivity variation of a fiber core and v is the fringe visibility associated with the index change. When two new coefficients are defined,

$$
\begin{aligned}
\sigma(z) &= \frac{\omega n_{co}}{2}\overline{\delta n_{co}}\int_{-\infty}^{\infty}\int_{-\infty}^{\infty} e_t(x,y) \cdot e_t^*(x,y) dxdy \\
\kappa(z) &= \frac{\omega n_{co}v}{4}\overline{\delta n_{co}}\int_{-\infty}^{\infty}\int_{-\infty}^{\infty} \Delta\varepsilon(x,y,z) e_t(x,y) \cdot e_t^*(x,y) dxdy
\end{aligned}
\tag{26}
$$

where σ is a dc coupling coefficient and κ is an ac coupling coefficient. By substituting Eqn. (25),(26) to Eqn. (24), the electric fields in the grating can be simplified to the superposition of forward-propagating and backward-propagating fundamental modes in the optical fiber core:

Fabrication and Mode Coupling of Long-Period Fiber Grating by Winding a Wire Around an Optical
Fiber Fixed to a Cylindrical Metal Rod

33

$$\frac{dR(z)}{dz} = i\hat{\sigma}(z)R(z) + i\kappa(z)S(z)$$

$$\frac{dS(z)}{dz} = -i\hat{\sigma}(z)S(z) - i\kappa^*(z)R(z),$$

(27)

where $R(z)=A^+(z)\exp[i(\delta z - \phi/2)]$ and $S(z)=A^-(z)\exp[-i(\delta z + \phi/2)]$. The general dc self-coupling coefficient can be represented by

$$\hat{\sigma} = \delta + \sigma - \frac{1}{2}\frac{d\phi}{dz} = \left(\beta - \frac{\pi}{\Lambda}\right) + \sigma - \frac{1}{2}\frac{d\phi}{dz},$$

(28)

where δ is the detuning factor, the ϕ and $d\phi/dz$ are the grating phase and possible chirp of the grating period, respectively. The solution of the complex reflection and transmission coefficient can be

$$R(z) = \frac{\gamma\cosh\left[\gamma\left(z - L/2\right)\right] + i\hat{\sigma}\sinh\left[\gamma\left(z - L/2\right)\right]}{\gamma\cosh\left(\gamma L\right) - i\hat{\sigma}\sinh\left(\gamma L\right)}$$

$$S(z) = \frac{-i\kappa\sinh\left[\gamma\left(z - L/2\right)\right]}{\gamma\cosh\left(\gamma L\right) - i\hat{\sigma}\sinh\left(\gamma L\right)}$$

(29)

where γ is described by

$$\gamma = \sqrt{\kappa^2 - \hat{\sigma}^2}, \ \left(\text{if, } \kappa^2 > \hat{\sigma}^2\right)$$

$$\gamma = i\sqrt{\hat{\sigma}^2 - \kappa^2}, \ \left(\text{if, } \kappa^2 < \hat{\sigma}^2\right),$$

(30)

The amplitude reflection coefficients can then be shown to be

$$\rho = \frac{S(-L/2)}{R(-L/2)} = \frac{-\kappa\sinh\left(\gamma L\right)}{\hat{\sigma}\sinh\left(\gamma L\right) + \gamma\cosh\left(\gamma L\right)},$$

(31)

2.3. Resonance wavelength

With the knowledge of the core and cladding effective indices, it is possible to determine LPFG resonance wavelengths for a specific grating period. This can be achieved by the phase matching condition in [16]. The transmission spectrum has dips at the wavelengths corresponding to resonances with various cladding modes. The resonance wavelength satisfying the phase-matching condition is given by [17]

$$\lambda_0 = \left(n_{eff,co}^{01} - n_{eff,cl}^{0m}\right)\Lambda,$$

(32)

where Λ is the period of the grating, and $n_{eff,co}^{01}$ and $n_{eff,cl}^{0m}$ are the effective indices of the LP$_{01}$ core mode and LP$_{0m}$ cladding modes at the resonance wavelength, respectively. Based on the phase-matched condition, the relationship between the period of the grating period (Λ)

and resonance wavelength (λ_0) can be obtained by the calculation of the propagation constants of the LP_{01} core and various cladding modes of a fiber at each specific wavelength. Figure 4 shows the resonance wavelength versus grating period characteristic for coupling with five orders of cladding modes, and for a specific grating pitch the spectrum yields resonance wavelength values for all visible bands.

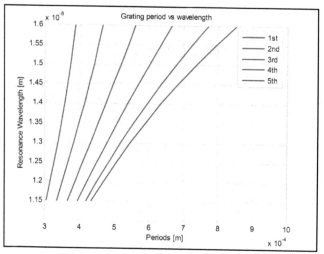

Figure 4. Resonance wavelength versus grating period.

Coupling mode theory applied to the three-layer model yields an expression for the core-cladding mode coupling coefficient by Erdogan [16]. The transmission spectrum modelled as Λ=455μm, L=20mm, and Δn_{co}=1×10⁻⁴ is shown in Fig. 5.

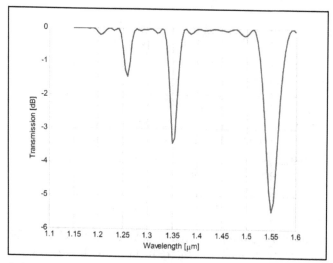

Figure 5. Transmission spectrum

3. Experiments and results

A LPFG has coupling characterisitics according to the phase-matching condition between a core mode and cladding modes in an optical fiber. A forward-propagating core mode (LP_{01} mode) is coupled to forward-propagating cladding modes (LP_{lm} mode) of the fiber. While LPFG generally allows coupling of a core mode to all order cladding modes (LP_{lm} mode), untilted WW-LPFG couples a core mode to only $l=0$ order cladding modes (LP_{0m} mode). The proposed WW-LPFG behaving as an un-tilted LPFG couples a core mode to LP_{0m} modes.

Figure 6 shows the structure and photo of the manufactured WW-LPFG. The SMF-28 standard fiber with a numerical aperture (NA) of 0.14, refractive index difference of 0.36 %, core and cladding diameters of 8.2 and 125 μm, respectively, was used. The refractive indices of core and cladding were 1.4489 and 1.444, respectively, at a 1550 nm wavelength. A copper wire with the diameter of 170 μm was used as a winding wire. The jacket of a SMF was stripped away to allow a copper wire to be wound around the fiber. The SMF was then attached to a cylindrical metal rod with a 10 mm diameter.

Figure 7 shows the apparatus and manufacturing process to fabricate the WW-LPFG. After an optical fiber was attached to the cylindrical metal rod, one end of a copper wire was glued to the cylindrical metal rod, and the other end of the copper wire was pulled by the mass of 900 g weight at the angle (θ) of 60 °, convertible to the tension of 8.8 N. When the cylindrical metal rod was rotated by a microprocessor, the wire holder controlled by a microprocessor is moved to the longitudinal direction of the cylindrical metal rod according to the designed pitch, so that the optical fiber was wound by the wire within the desired region. The lateral stress of the wire periodically pressing on the optical fiber changes the refractive index of the optical fiber, so the pitch of the winding wire is the key factor to the controlling of the resonance wavelength of the WW-LPFG. The experimental setup consisted of a broadband light source (Agilent, 83437A), the fabricated WW-LPFG, and an optical spectrum analyzer (Ando, AQ6315A). The light from the broadband light source propagated in the WW-LPFG, and the power intensity characteristics of the transmitted light were then determined when it reached the OSA.

Figure 8 shows the relationship of resonance wavelength with the grating period for the coupling of fundamental core mode to cladding modes. The transmission spectrum has dips at the wavelengths corresponding to resonances with various cladding modes. The resonance wavelength satisfying the phase-matching condition is given by Eqn. (32). Based on the phase-matched condition, the relationship between the period of the grating period and the resonance wavelength can be obtained by the calculation of the propagation constants of the LP_{01} core and various cladding modes of a fiber at each specific wavelength. Resonance wavelengths where guided-to-cladding mode coupling takes place can be theoretically determined by proper selection of a specific grating period. The vertical dashed line at the grating period of 500 μm meets with the wavelengths of 1280 nm, 1330 nm, 1380 nm, and 1490 nm of LP_{02}, LP_{03}, LP_{04}, LP_{05} modes, respectively. The resonance wavelengths of LP_{02}, LP_{03}, LP_{04}, LP_{05} modes correspond to 1280 nm, 1330 nm, 1380 nm, and 1490 nm, respectively. The parameters of the winding pitch, number of turns, and winding length of the copper wire are 500 μm, 40 turns, and 20 mm, respectively. Figure 9 shows the experimental and simulated transmission spectra of the WW-LPFG under the same conditions as those in Fig. 8. The simulated transmission spectra calculated by simulation software, OptiGrating (Optiwave Systems Inc) were in good agreement with the experimental result at the resonance wavelengths and transmission power loss. At 1490 nm, the transmission power gain of the resonance wavelength was -7.7 dB and the full width at half maximum (FWHM) was 20 nm.

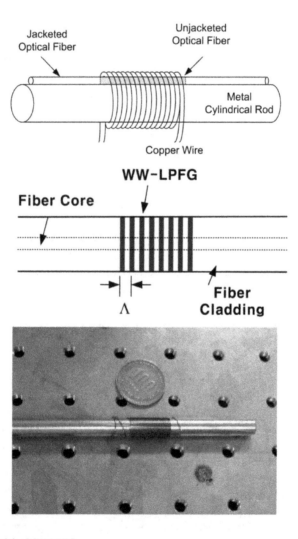

Figure 6. Structure of the WW-LPFG.

Fabrication and Mode Coupling of Long-Period Fiber Grating by Winding a Wire Around an Optical
Fiber Fixed to a Cylindrical Metal Rod

37

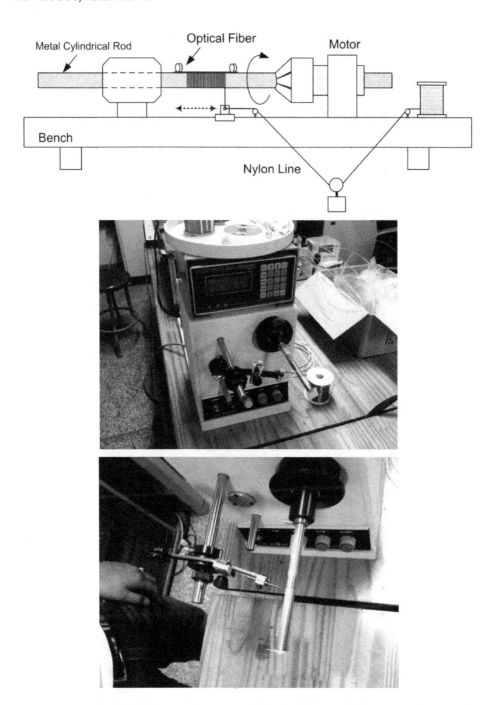

Figure 7. Experimental setup and apparatus

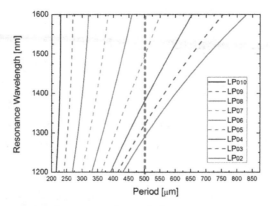

Figure 8. Resonance wavelength versus grating period.

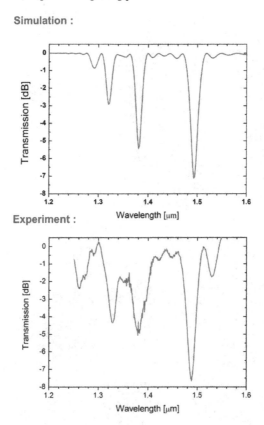

Figure 9. Transmission spectrum at the grating pitch of 500um

The spectral dependence of the power transmission of the LP_{01} mode in the LPFG can be determined by [16]

Fabrication and Mode Coupling of Long-Period Fiber Grating by Winding a Wire Around an Optical
Fiber Fixed to a Cylindrical Metal Rod

39

$$\frac{P_{01}(L)}{P_{01}(0)} = \cos^2\left(\sqrt{\kappa^2 + \delta^2}\,L\right) + \frac{1}{1 + \dfrac{\kappa^2}{\delta^2}}\sin^2\left(\sqrt{\kappa^2 + \delta^2}\,L\right), \tag{33}$$

where δ is the detuning parameter

$$\delta = \frac{1}{2}\left\{\beta_{co}^{01} - \beta_{cl}^{0m} - \frac{2\pi}{\Lambda}\right\}, \tag{34}$$

κ is the coupling constant for the grating, L is the grating length, and β_{co}^{01} and β_{cl}^{0m} are the propagation constants of the LP_{01} core mode and LP_{0m} cladding modes, respectively. The standard silica optical fiber is a cylindrical and isotropic dielectric waveguide with a circularly symmetric refractive index distribution. However, as the applied tension of the wire induces deformations of the optical fiber, the refractive indices parallel and perpendicular to the direction of the applied tension are differently changed and the birefringence is derived from the photo-elastic property of the fiber. The refractive index changes at any point of the fiber are [18]

for x-polarization,

$$\left(\Delta n_{eff}\right)_x = -\frac{1}{2}\left(n_{eff,0}\right)^3\left[P_{11}\varepsilon_x + P_{12}\left(\varepsilon_y + \varepsilon_z\right)\right], \tag{35}$$

for y-polarization,

$$\left(\Delta n_{eff}\right)_y = -\frac{1}{2}\left(n_{eff,0}\right)^3\left[P_{11}\varepsilon_y + P_{12}\left(\varepsilon_x + \varepsilon_z\right)\right]. \tag{36}$$

Here, $n_{eff,0}$ is the initial effective refractive index of the core of unperturbed fiber, P_{11} and P_{12} are the photo-elastic coefficients of the unperturbed optical fiber, and ε_x, ε_y, ε_z are the strain components at the fiber in the x, y, and z direction, respectively. Because of the mechanical and photo-elastic properties of the optical fiber's material ($P_{11} < P_{12}$, $\varepsilon_x > 0$, $\varepsilon_y < 0$, $\varepsilon_x < \varepsilon_y$), the refractive index change of the x-polarization is higher than that of the y-polarization. Since the circular fiber of the fabricated WW-LPFG is periodically pressed in one direction to form a mechanically induced long-period grating, birefringence due to the asymmetric structure occurs.

As shown in Fig. 10, the spectrum of the fabricated WW-LPFG has polarization dependence on x- and y-polarization lights. When an LP_{01} core mode is coupled to an LP_{04} cladding mode at a 520 µm grating pitch, the resonant wavelengths of x- and y-polarization light appeared at 1435 and 1426 nm, respectively. By the birefringence effect, the polarization dependence makes the change of 9 nm resonance wavelength and 2.5 dB transmitted light power. Figure 11 shows the transmission spectra of the LP_{04} mode of the WW-LPFG with the pitch of 500 µm and with several masses of 0.6, 0.9, and 1.2 kg weight, convertible to the tension of 5.8, 8.8, and 11.7 N, respectively. The higher tension by the winding wire pressure induced the higher index variation, which resulted in stronger mode coupling and thus generally in deeper dips in the transmission spectra. Meanwhile, the resonant wavelength did not shift by the change of the applied tension.

The transmission spectra of the LP04 mode of the WW-LPFG with the diameters of the cylindrical metal rod of 7 and 10 mm were shown in Fig. 12 at the fixed pitch of 500 μm. When the diameter of the cylindrical metal rod is smaller, the pressing region on the fiber attached to the cylindrical metal rod is wider, and thus the higher index variation of the fiber results in deeper dips in the transmission spectrum.

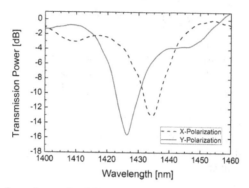

Figure 10. Polarization dependence of an LP04 mode at a grating pitch of 520um

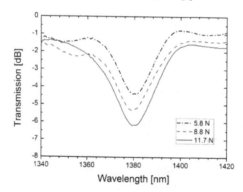

Figure 11. Transmission spectra of the WW-LPFG with different wire tensions.

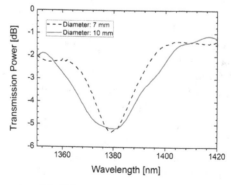

Figure 12. Transmission spectra of the WW-LPFG with different diameters of the cylindrical metal rod.

4. Conclusion

An easily fabricated WW-LPFG is wire-wound periodically around an optical fiber attached to a cylindrical rod. The periodic pressure of the wire results in refractive index change of an optical fiber core, and the resonance wavelengths of the grating can be easily controlled according to the winding-wire pitch controlled by a microprocessor. Due to the simple and flexible fabrication process using the winding of a wire, the fabricated WW-LPFG is small, and cost-effective, and the resonance wavelength are easily controlled by the pitch of the winding-wire using a microprocessor. At the winding-wire pitch of 500 μm and 510 μm, the spectra show resonance-wavelength dips corresponding to the LP_{02}, LP_{03}, LP_{04}, LP_{05} cladding modes. The polarization dependence of a resonance wavelength shift of 9 nm was shown. Smaller diameter of the cylindrical metal rod and higher tension of the winding wire led to the stronger mode coupling according to induced higher index variation. With smaller diameter of the cylindrical metal rod, or higher tension due to the winding pressure, stronger mode coupling can result in deeper dips in the transmission spectra according to induced higher index variation.

Author details

Jonghun Lee and Cherl-Hee Lee
*Robotics Research Division, Daegu Gyeongbuk Institute of Science & Technology,
Daegy, Korea*

Kwang Taek Kim
Department of Optoelectronics, Honam University, Gwangju, Korea

Jaehee Park
Department of Electronic Engineering, Keimyung University, Daegu, Korea

Acknowledgement

This work was supported by the DGIST R&D Program of the Ministry of Education, Science and Technology of Korea(12-RS-02).

5. References

[1] Vengsarkar A M, Lemaire J P, Judkins J B, Bhatia V, Erdogan T, Sipe J E. Long-period Fiber Gratings as Band-rejection Filters. Journal of Lightwave Technology 1996; 14: 58-65.

[2] Ostling D, Engan H E. Broadband Spatial Mode Conversion by Chirped Fiber Bending. Optics Letter 1996; 21: 192-194.

[3] Vengsarkar A M, Pedrazzani J R, Judkins J B, Lemaire J P, Bergano N S, Davidson C R. Long Period Fiber Grating Based Gain Equalizers. Optics Letter 1996; 21: 336-338.

[4] Bhatia V, Vengsarkar A M. Optical Fiber Long Period Grating Sensors. Optics Letter 1996; 21: 692-694.

[5]   Martin J, Ouellette F. Novel Written Technique of Long and Highly Reflection In-fiber Gratings. Electron Letters 1994; 30: 811-812.

[6] Humbert G, Malki A. Annealing Time Dependence at Very High Temperature of Electric Arc-induced Long-period Fibre Gratings. Electron Letters 2002; 38: 449-450.

[7] Bae J K, Kim S H, Kim J H, Bae J, Lee S B, Jeong J M. Spectral Shape Tunable Band-rejection Filter Using a Long-period Fiber Grating with Divided Coil Heaters. IEEE Photonics Technology Letters 2003; 15: 407-409.

[8] Enomoto T, Shigchara M, Ishikawa S, Danzuka T, Kanamori H. Long-period Grating in a Pure-silica-core Fiber Written by Residual Stress Relaxation. OFC'98 1998; 277-278.

[9] Savin S, Digonnet M J F, Kino G S, Shaw H J. Tunable Mechanically Induced Long-period Fiber Gratings. Optics Letter 2000; 25: 710-712.

[10] Lee N K, Song J W, Park J H. Fabrication of Fiber Device with Long-period Fiber Gratings at Locations Under Applied Pressure and its Application as Load Sensor. Japanese Journal of Applied Physics 2006; 45: 1656-1657.

[11] Wu E, Yang R C, San K C, Lin C H, Alhassen F, Lee H P, A Highly Efficient Thermally Controlled Loss-tunable Long-period Fiber Grating on Corrugated Metal Substrate. IEEE Photonics Technology Letters 2005; 17: 612-614.

[12] Lee N K, Song J W, Park J H. Mechanically Induced Long-period Fiber Grating Array Sensor 2011; 53: 2295-2298.

[13] Rego G, Fernandes J R A, Santos J L, Salgado H M, Marques P V S. New Technique to Mechanically Induce Long-period Fibre Gratings, Optical Communications 2003; 220: 111-118.

[14] Rego G. Long-period Fiber Gratings Mechanically Induced by Winding a String Around a Fiber/grooved Tube Set, Microwave and Optical Technology Letters 2008; 50: 2064-2068.

[15] Lee C-H, Lee J, Park J, Kim K T. Easy Fabrication of Long-period Fiber Grating by Winding a Wire around an Optical Fiber Fixed to Cylindrical Rod. Microwave and Optical Technology Letters 2012; 54: 1937-1941.

[16] Erdogan T. Cladding-mode Resonances in Short- and Long-period Fiber Grating Filters. Journal of Optical Soceity of America A 1997; 8: 1760-1773.

[17] Erdogan T. Fiber Grating Spectra. Journal of Lightwave Technology 1997; 15: 1277-1294.

[18] Gafsi R, El-Sherif M A. Analysis of Induced-birefringence Effects on Fiber Bragg Gratings. Optical Fiber Technology 2000; 6: 299-323.

Chapter 3

Surface-Corrugated Microfiber Bragg Grating

Fei Xu, Jun-Long Kou, Yan-Qing Lu, Ming Ding and Gilberto Brambilla

Additional information is available at the end of the chapter

1. Introduction

Optical fiber Bragg gratings (FBG) are key devices for optical communication and sensors. FBGs are based on a periodic variation in the refractive index of the fiber core and allow for the reflection of a narrowband signal centered at a specific Bragg wavelength. Over the last two decades, FBGs have been manufactured mainly by modifying the core refractive index using interferometric or point-by-point techniques; most of interferometric techniques use a phase mask and an ultraviolet (UV) laser [1] (typically excimer or frequency doubled Ar+ ion) or femtosecond (fs) lasers (near IR [2] or UV [3]). Gratings based on surface corrugations have also been demonstrated in etched fibers using photolithographic techniques [4]. All gratings fabricated in thick fibers have weak refractive index modulations ($\Delta n_{mod} \sim 10^{-4}$-$10^{-3}$) and the related grating lengths are of the order of several millimeters.

Structural miniaturization is one of the current trends for achieving higher-bandwidth, faster response and higher-sensitivity. Experimentally, FBG miniaturization has been achieved in two steps. Firstly, the fiber has been tapered into a subwavelength-scale microfiber (MF), considered to be the basic element for miniature fiberized devices and subsystems [5]. Then strong refractive index modulations ($\Delta n_{mod} > 10^{-1}$) are induced. Large Δn_{mod} can be obtained by alternating layers of different materials, one of which can be air. Although this process in normal optical fibers imposes the removal of large amounts of material (the propagating mode is confined at a depth >50 μm from the fiber surface), in fiber tapers and tips it only requires the removal of ~10,000 less matter because the propagating mode is confined by the silica/air interface in areas with micron size. A few techniques have been proposed to fabricate surface-corrugated fiber gratings (SCMGs), including photorefractive inscription using etching [6], femtosecond lasers [7, 8], and focused ion beam (FIB) [9-16]. So far, FIB is the most flexible and powerful tool for patterning, cross-sectioning or functionalizing a subwavelength circular microfiber due to its small and controllable spot size and high beam current density. In the past two years, a number of ultra-compact SCMGs have been successfully fabricated by FIB milling, with lengths as small as ten micrometers. In addition, a variety of other techniques have been proposed to generate gratings exploiting the fraction of power propagating in the evanescent field: these include wrapping a microfiber on a microstructured rod or put a microfiber on a surface-corrugated planar grating. This chapter reviews recent advances in ultra-small SCMGs, their characteristics and applications.

2. Theory

2.1. SCMG spectral properties

In SCMGs, strong Δn_{mod} is achieved alternating layers of air and glass. Figure 1 shows a typical SCMG geometry: the surface-corrugated structure is fabricated by inducing periodic microscale open-notches on the side or holes across the optical microfiber.

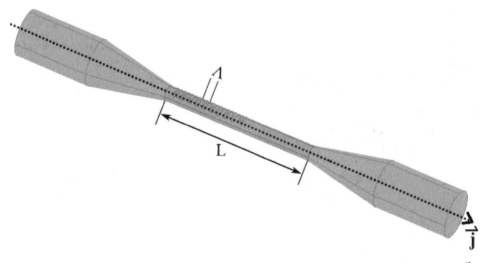

Figure 1. Schematic illustration of SCMGs. L and Λ are the grating length and period, respectively. $\vec{j}$ is the unit vector along the fibre longitudinal axis.

In the grating, the forward (f) and backward (b) propagating modes are related by

$$\vec{\beta_b} = \vec{\beta_f} + m\frac{2\pi}{\Lambda}\vec{j}. \tag{1}$$

where $\beta_i = (2\pi/\lambda)n_{eff,i}$ (i = f or b) is the mode propagation constant, m is the diffraction order, Λ is the period of the grating and $\vec{j}$ is the unit vector along the fibre longitudinal axis. In a more physical perspective, Eq. (1) means that the momentum mismatch between the forward and backward propagating modes should be compensated by the reciprocal vector provided by the periodical index modulation. For the first-order diffraction which is commonly seen in SCMGs:

$$\left(n_{eff,f} + n_{eff,b}\right)\Lambda = \lambda_B. \tag{2}$$

Furthermore, if the two modes are identical, the commonly used Bragg resonance condition can be obtained, namely, $\Lambda_B = 2n_{eff}\Lambda$.

For a SCMG, the effective index difference between the MF milled and un-milled cross sections can be as large as $\sim 10^{-3}$[9], or even $\sim 10^{-1}$[12], orders of magnitude larger than that in conventional FBGs. One way to calculate an averaged effective index of the grating region is to choose an unperturbed waveguide boundary using the method developed by W. Streifer [17]: the boundary between different materials is shifted to compensate for different geometry, shown as the dashed line in Fig. 2. d is the depth of the corrugation and h_{eff} is the boundary shift from the top of the corrugation to the new boundary of the corresponding unperturbed waveguide. The boundary shift h_{eff} illustrated in Fig. 2(b) and (c) is determined so that the volume bounded by the upper part of the corrugation ($S_A\tau$) is equal to the volume bounded below [$S_B(1-\tau)$], i.e.

$$S_A\tau = S_B(1-\tau). \tag{3}$$

where τ is the duty cycle. The averaged effective index could thus be obtained by mode analysis after the unperturbed waveguide boundary is established.

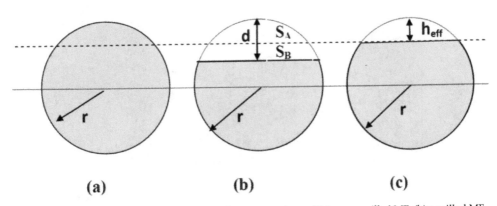

(a)   **(b)**   **(c)**

Figure 2. Equivalent unperturbed geometry. The cross-sections of (a) an un-milled MF, (b) a milled MF and (c) an equivalent unperturbed geometry, respectively. d is the groove height and h_{eff} is effective height, obtained solving eq. 3..

The SCMG reflection spectrum can be estimated using [18]. However, due to the large index difference in the corrugation region, strong scattering may occur. A more effective way to verify the experimental spectrum obtained from the SCMG is a 3D finite element simulation, as shown in Fig. 3. This method takes the details of the structural deformation into consideration and thus better reflects the real situation experienced by the light field.

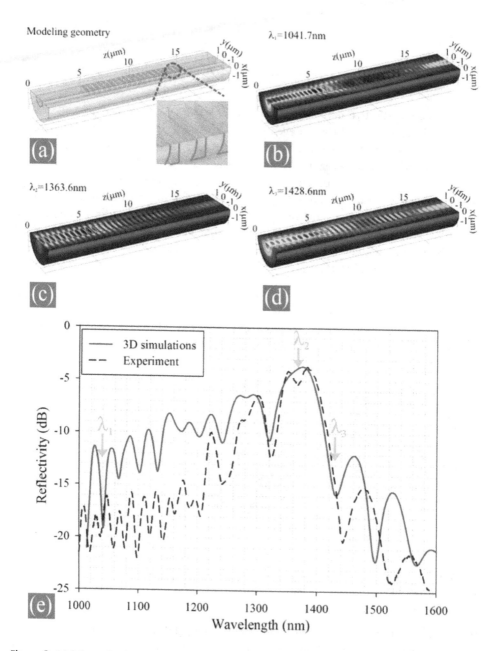

Figure 3. (a) Schematic of a 3D finite element simulation of SCMG; the insert shows the magnified figure of the biconcave air notch. (b), (c), and (d) Electric fields at wavelength $\lambda_1 = 1041.7$nm, $\lambda_2 = 1363.6$nm, and $\lambda_3 = 1428.6$nm, respectively. (e) SCMG reflection spectra. The red solid line is the 3D simulation line while the blue dashed line is the experiment result. λ_1, λ_2, and λ_3 represent the wavelengths whose electric fields are shown in (b), (c), and (d) [19].

2.2. SCMG sensing mechanisms

In the past years, FBGs have been widely applied to the measurement of chemical, biomedical, physical, electrical parameters, especially for structural health monitoring in civil infrastructures, where the information of measurands usually rely on the monitoring of the shift in wavelength of the reflected Bragg signal with the changes in the measurand (*e.g.*, refractive index, temperature, and strain/force). FBG based sensors have a number of advantages with respect to conventional sensors, such as compactness, immunity to electromagnetic interference, multiplexing, rapid response to real time monitoring, and high sensitivity to external perturbations. Compared with conventional FBGs, SCMGs have similar applications and advantages with some extra benefits such as ultra-compact size and large evanescent field.

As the effective index and period of grating is a function of r, n_a, T, and ε, the Bragg condition (eq. 2) can be rewritten as [19]

$$\lambda_B = 2n_{eff}(r , n_f, n_a, T, \varepsilon)\Lambda(T, \varepsilon). \tag{4}$$

where the refractive index of fused silica and of the ambient medium surrounding the SCMGs are denoted by n_f and n_a, T is the temperature and ε is the strain applied to the SCMGs.

2.2.1. Refractive index

The notable distinction between SCMGs and conventional FBGs lies in the large SCMG evanescent field which enables its capabilities for external medium sensing. When a SCMG is operated as a refractive index (RI) sensor, the wavelength shift depends on the change of n_a. The sensor sensitivity (S_a) with respect to the ambient medium RI is defined as [19]

$$S_a = \frac{d\lambda_B}{dn_a} = \frac{\partial \lambda_B}{\partial n_{eff}(n_a, r_{MF})} \frac{\partial n_{eff}(r_{MF}, n_a)}{\partial n_a} = 2\Lambda \frac{\partial n_{eff}}{\partial n_a}. \tag{5}$$

Figure 4 shows S_a as a function of r_{MF}. n_a is chosen to be 1.33 and 1.42, because most of the RI sensors work around these values. S_a increases for decreasing MF diameters, as larger fractions of power propagate in the evanescent field, thus in the surrounding environment. For the same reason, higher external RIs are associated with a larger S_a. Theoretically, the largest S_a which can be obtained by SCMG is ~2Λ, typically around 1100 nm/RIU according to Eq. (5). This value is comparable to optical microfiber coil resonator sensor [20, 21] and higher than sensors based on microcapillary resonator [22] or photonic crystal microresonator [23].

2.2.2. Temperature

Temperature affects the Bragg wavelength shift through the thermo-optical effect and thermal expansion in three ways: index variation, MF radius variation and grating period change, each of which is represented in Eq. (6). The temperature sensitivity (S_T) can be defined as [19]:

$$S_T = \frac{d\lambda_B}{dT} = 2\Lambda\left(\sigma_T \frac{\partial n_{eff}}{\partial n_f} + r\alpha_T \frac{\partial n_{eff}}{\partial r} + n_{eff}\alpha_T\right) \tag{6}$$

Here, σ_T (1.2×10^{-5}/°C) is the thermo-optical coefficient and α_T (5.5×10^{-7}/°C) is the thermal expansion coefficient of fused silica. As thermal expansion contributes less than 2 pm/°C to the total sensitivity, it is generally neglected. S_T resulting from the thermo-optical effect is ~ 10 – 20 pm/°C and dominates in temperature sensing, which is in agreement with previous results obtained using fiber tip Fabry-Perot interferometers [24].

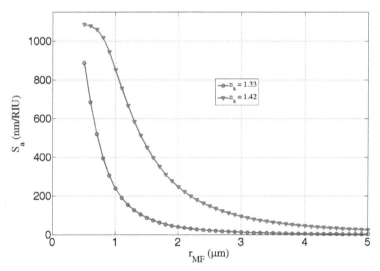

Figure 4. Dependence of sensor sensitivity S_a on the MF radius r_{MF} in the range 0.5 – 5μm for two values of surrounding medium refractive indices n_a. Resonant wavelength is set at 1550nm and n_a is chosen to be at 1.33 (blue circles) and 1.42 (red triangles). Only the fundamental mode is considered.

2.2.3. Strain/Force

From continuum mechanics, when longitudinal strain is applied to a SCMG, its Bragg wavelength shift can be estimated as follows [19]:

$$\Delta\lambda_B = 2\Lambda n_{eff}\left\{1 - \left(\frac{n_{eff}^2}{2}\right)\left[p_{12} - v(p_{11} + p_{12})\right]\right\}\varepsilon = 2\Lambda n_{eff}\left(1 - p_{eff}\right)\varepsilon, \tag{7}$$

where ε is the applied strain, v is Poisson's ratio and p_{ij} coefficients are the Pockels' strain-optical tensor coefficients of the fiber material. If the MF cross section deformation due to the applied strain is neglected, strain sensitivity (S_s) is reduced to

$$S_S = \frac{\Delta \lambda_B}{\varepsilon} = \lambda_B \left(1 - p_{eff}\right). \tag{8}$$

The effective photo-elastic coefficient p_{eff} for a SCMG strain sensor is ~ 0.21, giving S_S ~ 1.2 pm/$\mu\varepsilon$, which is comparable to that of a conventional fiber grating with a Bragg wavelength of 1550 nm. This is in agreement with experimental results [25, 26]. SCMG sensors can also be characterized by their force sensitivity (S_F)

$$S_F = \frac{S_S}{\pi r_{MF}^2 E}, \tag{9}$$

where E represents the fibre Young's modulus. Equation (9) shows that S_F scales inversely with the square of the microfiber diameter.

For composite structures, like the SCMG wrapped on a PCF mentioned below, sensitivity depends on more parameters such as the strain-optical tensor of the photonic crystal fiber (PCF) support and the embedding coating.

3. SCMG fabrication

Several techniques have been reported in the literature for the fabrication of gratings in MFs and they can be classified as follows [19]:

1.   Etch-eroded commercial FBG or UV irradiated FBG [25-33].
2.   FIB-milled FBG on MFs [9-16].
3.   Femtosecond-laser-irradiated FBG on MFs [7, 8].
4.   FBGs exploiting evanescent fields [34-37].

Both cladding-etched commercial FBGs and UV irradiated FBGs in MFs are uniform microfiber gratings, meaning that the grating region experiences only RI modulation and not structural perturbations as indicated in Fig. 1. The most commonly used technique to get an uniform microfiber grating is to etch a single mode fiber (SMF) after the FBG has been written in the photosensitive Ge-doped core [25, 27-29]. Usually, a hydrofluoric acid aqueous solution (~ 20% – 50%) at room temperature is employed for the etching process at an etching speed of ~ 0.5 – 2 μm/min. The diameter of the etched fiber can be measured and controlled in situ by monitoring the transmission loss.

Femtosecond-laser-irradiation is another way to induce periodical physical deformations on the surface of MFs [8]. During the femtosecond laser irradiation, the ultra-short laser pulses transfer energy to the electrons in the material irradiated through nonlinear ionization [38]. When a sufficiently high energy is achieved, pressure or shock waves cause melting or non-thermal ionic motion, resulting in permanent structural damages in the material. Aided with proper phase masks, uniform microfiber gratings can be fabricated on the surface of the MFs [7, 8].

For the fabrication of ultra-small SCMGs, the main technique is FIB milling or inducing external gratings.

3.1. FIB-milled SCMGs

FIB milling, a powerful micromachining technique, has been the tool of choice to fabricate SCMGs [9-16]. This method employs accelerated ions to mill nanometer-scale features on MF surfaces to form corrugated structures. As the index modulation results from changes in the structure, this kind of gratings are called structural SCMGs.

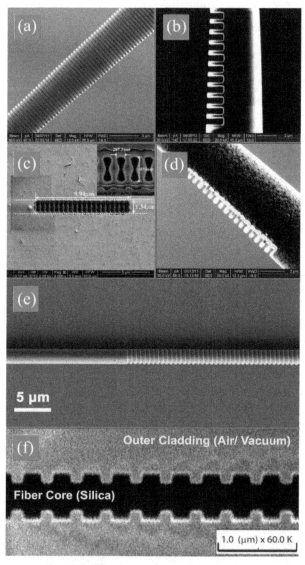

Figure 5. FIB/SEM pictures of gratings fabricated on MF tips (a)[9], (b)[10], (d)[13] and MF tapers (c)[12], (e)[14], (f)[15]. Reprinted with permission. Copyright 2011 Optical Society of America, Copyright 2012 AIP and Copyright 2011 IEEE.

Prior to milling, the MF is coated with a thin film of metal, *e.g.* aluminum or gold [9-13], to prevent charge accumulation which cause ion deflections and large fabrication errors. Alternatively, MF can also be laid on a doped silicon wafer [14]: due to van der Waals' forces, the MF tightly attaches to the conductive substrate and it avoids charging by transferring charges to the wafer.

During the FIB micromachining process, the MF sample should be fixed firmly in the vacuum chamber to minimize sample displacements. A 30kV, 10 – 300pA Ga$^+$ ion beam is usually used to get a good milling accuracy. The total milling process takes minutes to hours according to the beam current used and milled area. After the machining process, when metal coating is not required, the MF is immersed in metal etchant to totally remove the metal film and then is cleaned with deionized water.

Figure 5 shows FIB/SEM pictures of SCMGs fabricated from different groups. Gratings in Fig. 5(a), (b), and (d) are fabricated on MF tips while the rest are on tapers. SCMGs are fabricated on MFs with diameter ranging from 560 nm [Fig. 5(f)] to 6.6 μm [Fig. 5(a)] and the number of grating periods varies from 11 [Fig. 5(d)] to 900 [Fig. 5(e)]. Both high [5×10^{-3} – 10^{-1}, Fig. 5(c), (d), (f)] and low [10^{-4} – 5×10^{-3}, Fig. 5(a), (b), (e)] average RI modulations have been achieved by FIB-milled SCMGs. In all, FIB provides researchers a flexible way to get all kinds of structures with high accuracy at will and without additional masks. Yet, batch production cannot be envisaged for this method.

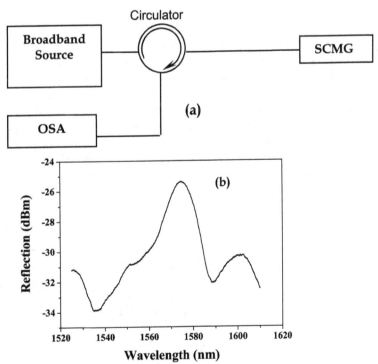

Figure 6. (a) Experimental set-up of characterizing SCMGs. (b) Reflection spectra of a SCMG in air.

Experimentally, SCMGs are characterized with a set-up similar to that shown in Fig. 6(a): light from a broadband source is injected into the grating and the reflected light is then collected by an optical spectrum analyzer (OSA) (Ando AQ6317B, Japan) after passing through a circulator. The fiber tip reflection has been analyzed before grating inscription and displayed a negligible reflection over the whole spectrum, showing that the reflection at the tip end was insignificant. Fig. 6(b) shows the typical reflection spectrum of a SCMG. There is a clear peak at ~ 1572 nm with an extinction of 10 dB. Two side lobes are observed at longer (1600 nm) and shorter (1535 nm) wavelengths. The possible explanation of the wide bandwidth of the experiment reflection spectrum can be the effective indices variation, rough notch surfaces and departure from perfect grating periodicity.

3.2. SCMG exploiting evanescent fields

In addition to the previous techniques, other methods have demonstrated or proposed the fabrication SCMGs using the strong microfiber evanescent field. SCMGs can be manufactured by wrapping a MF on a microstructured rod and then coating them with a low-loss polymer. The rod can be obtained by etching a cane used in the manufacture of microstructured fibers or it can be a thick microstructured fiber. Here no expensive mask, laser or machining set-up is used. Combining current enabling technologies on microstructured optical fibers [39, 40] and MFs [41], it is possible to induce periodic or variable corrugations leading to the coupling between forward and backward propagating waves. The final SCMG structure is shown in Fig. 7(a) [34, 37]: it is a compact and strong micro-device with some air holes arranged in a circle. Fig. 7(b) presents a possible rod cross-section. The mode propagating in the microfiber experiences refractive index corrugations because its evanescent field overlaps with the support rod where there is alternation of glass and air holes. Here we only consider one ring of air holes; the inner large hole is also filled with air and the MF is assumed perpendicular to the rod. When a coordinate increasing along the microfiber is used, the surface corrugations experienced by the mode propagating along the curvilinear coordinate are similar to those experienced by a mode propagating straight in proximity of a conventional planar grating. Unfolding the MF, the MF grating can be taken as a coated MF on a planar substrate with air-hole corrugations: the equivalent structure is shown in Fig. 7(c) [34, 37].

The mode field in the unperturbed waveguide geometry is derived from the perturbed (straight) geometry using the method mentioned in section 2.1. If g_1 is the boundary between the materials with refractive indices n_1 (air) and n_2 (the rod material, generally silica) and g_2 is the boundary between the materials with refractive indices n_4 (coating) and n_2, a new g_0 is chosen to match the volume of n_1 material extending into region A (above g_0) with the volume of n_2 material extending into region B (below g_0), as shown in Fig. 8(a). We introduce the effective distance d_{eff} between g_2 and g_0 as[34, 37]:

$$d_{eff} = g_1 + g_2 - g_0 = d_1 + d_2 + 2r - \pi r^2 / \Lambda \qquad (10)$$

d_{eff} increases for increasing $d = d_1 + d_2$ and increasing Λ.

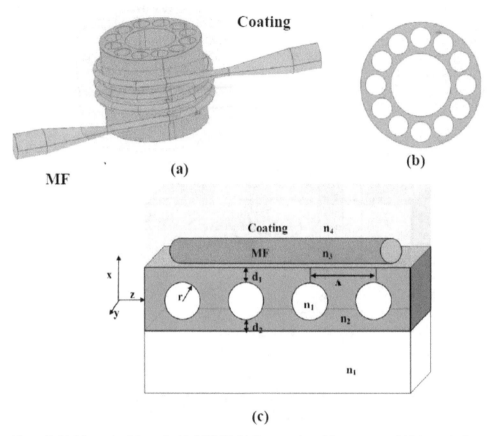

Figure 7. (a) Schematic of the embedded SCMG. (b) Cross-section of the support rod (c) the equivalent planar structure, n_1, n_2, n_3, n_4 are the refractive indices of the hole, the rod, the MF, and the coating, respectively[34, 37]. Copyright 2009 Optical Society of America and Copyright 2010 IEEE.

This method is also extremely flexible and can be used to manufacture chirped MF gratings: chirping can be achieved by tuning the hole radius and pitch, or by changing the distance between MF and holes. Since $\lambda = 2n_{neff}(d_{eff})\Lambda / m$, there are three simple ways to realize chirp: by tuning Λ, $d = d_1 + d_2$, and r. Chirp rates are defined as [34, 37]:

$$
\begin{cases}
\dfrac{\partial \lambda}{\partial \Lambda} = \dfrac{2}{m}\dfrac{\partial n_{neff}}{\partial d_{eff}}\dfrac{\pi r^2}{\Lambda} + \dfrac{2}{m}n_{neff} \\[2mm]
\dfrac{\partial \lambda}{\partial r} = \dfrac{2}{m}\dfrac{\partial n_{neff}}{\partial d_{eff}}(2\Lambda - 2\pi r) \\[2mm]
\dfrac{\partial \lambda}{\partial d} = \dfrac{2}{m}\dfrac{\partial n_{neff}}{\partial d_{eff}}\Lambda
\end{cases}
\tag{11}
$$

All the derivatives $\partial\lambda/\partial\Lambda$, $\partial\lambda/\partial d$ and $\partial\lambda/\partial r$ depend on $\partial n_{neff}/\partial d_{eff}$, which in turn is strongly dependent on d_{eff} and r. $\partial n_{neff}/\partial d_{eff}$ is of the order of 10^{-2} μm^{-1} and can achieve the maximum at $d_{eff} \sim 1$ μm. For m = 2, from Eq. (11) $\partial\lambda/\partial\Lambda \gg \partial\lambda/\partial d \sim \partial\lambda/\partial r$; since $\partial\lambda/\partial\Lambda > 1$, chirped grating can be easily realized by tuning the grating period: current PCF fabrication techniques make the low-cost period-tuning possible. Moreover, $\partial\lambda/\partial d \sim \partial\lambda/\partial r \sim 10^{-5}$: which is enough to produce small precise chirps in a simpler and cheaper way than inducing a temperature, strain or refractive-index gradients [34, 37].

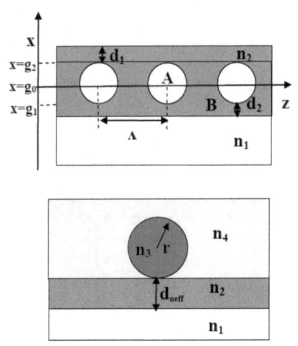

Figure 8. Top: Illustration of the outer layer structure of the support rod in Streifer's theory. n_1 and n_2 represent the refractive indices of the hole and rod, Λ the distance between two adjacent holes, g_1 and g_2 the boundaries between different layers and g_0 represents the new equivalent boundary between n_2 and n_1. Bottom: cross-section of the MF in the equivalent outer straight layer structure; n_3 and n_4 are the indices of the MF and coating; r is the radius of the MF and d_{eff} is the effective wall thickness[34, 37]. Copyright 2010 Optical Society of America and Copyright 2009 IEEE.

These methods to induce chirp mainly depend on the geometry of the support rod; in fact hole pattern has a periodicity related to the MF turn around the support rod, the circumference of which limits the chirp length. An alternative method to achieve chirped gratings relies on tilting the MF with respect to the support rod longitudinal axis: if the MF cross the rod at an angle ϕ the Bragg condition is expressed by [42] $\lambda = 2n_{eff}\Lambda / \cos\varphi / m$; by changing ϕ gradually, the Bragg wavelength changes gradually, it is possible to control the grating Bragg wavelength and get a chirped grating. When ϕ is very small, the chirp rate is given by [34, 37]

$$\frac{\partial \lambda}{\partial \varphi} = -2\frac{n_{eff}\Lambda \sin\varphi}{m\cos^2\varphi} \approx 2\frac{n_{eff}\Lambda\varphi}{m} = \lambda\varphi \quad (12)$$

The chirp range varies considerably: from 0.3 nm/° at φ = 0.1° to 300 nm/° at φ = 10°. These two chirping methods allow for a great deal of flexibility in the chirp design[34, 37].

The use of microstructured support rods to make gratings provides an extreme flexibility because it is easier to be dealt with than the microfiber itself. By controlling the air holes geometry, it is possible to get several-layer corrugations or phase shifted gratings. If the rod is coated with an active layer or the holes are filled with an active medium, it generates a laser. If holes are used as microfluidic channels, the support rod can work as a sensor. Moreover, since only a very short piece of rod is needed for each device, the average device cost is very low. The device can be coated with stable polymer such as Teflon and UV375 and keep it very strong [34, 37].

A similar method relies on laying the MF on a substrate with pre-treated microstructures [see Fig. 9]. The fraction of power propagating in the evanescent field interacts with the periodically distributed patterns in the rod or the substrate, and light transmission can then be modulated. Both methods avoid post-processing the thin MFs and have great flexibility. However, the MFs have to be coated with low index polymer [43] which means that they are not suitable for high temperature sensing.

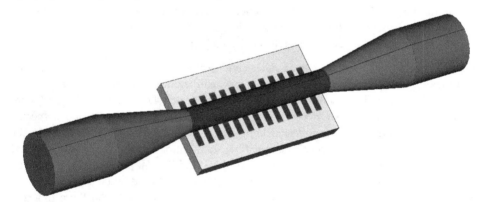

Figure 9. Proposed SCMG by laying the MF on a substrate with pre-treated microstructures[44].

Finally, Ding *et al.*[35] combined metal lift-off technology with lithography to produce metallic surface gratings, which provided a high and constant sensitivity to the ambient medium RI, while Phan Huy *et al.* [36] demonstrated an improvement in the sensitivity of RI by making use of the suspended core of a microstructured fiber.

4. Sensing applications

SCMGs can have a number of possible sensing applications as conventional FBGs. Up to date, SCMGs have been used to measure refractive index, temperature and strain/force.

With the extreme small size and flexible geometry, SCMGs offer great prospects for developing novel sensors with a very small perturbation on the object being measured.

4.1. Refractive index sensing

Much of the SCMG applications relate to RI sensing because of the very large MF evanescent field. For a typical SCMG sensor immersed in ambient liquid with RI in the range 1.32 – 1.46, S_a varies from 10 nm/RIU (refractive index unit) to 10^3 nm/RIU, according to the MF radius and the ambient liquid sensed, whatever for a FIB-milled or rod-wrapped SCMG. Usually, a smaller radius and a larger ambient medium RI result in a higher sensitivity regardless of the fabrication method. For example, Liang *et al.* got a sensitivity of 16 nm/RIU at a RI around 1.35 with a MF 6 µm in diameter [27] while 660 nm/RIU was reached by Liu *et al.* at a RI of 1.39 by using a 1.8 µm-diameter MF [14]. Both of them agree well with what is predicted from Eq. (5).

In addition to all these nonmetallic SCMGs, metallic gratings have also been proposed for RI sensing. The existence of metal causes light to be coupled to modes of different properties [10, 35]. Figure 10(a) shows the reflection spectra of the metal-dielectric-hybrid SCMG [Fig. 5(d)] immersed in air, acetone, and isopropanol, respectively. The extinction ratio is about ~ 10 dB. The small degree of chirp has been ascribed to the non-uniformity of the taper. The taper diameter difference at the grating extremities is less than 1 µm as illustrated in Fig. 5(d). According to simulations, different effective refractive indices resulting from the variation of the diameter induce resonant wavelength shifts of ~ 4 nm. Some small ripples in Fig. 10(a) can probably be due to a degree of chirping [10].

The grating spectrum presents several valleys and peaks with different characteristics in a 100 nm wide spectral range [10]. The peaks shift when the outer environment changes from acetone to isopropanol. However, these valleys and peaks show larger shifts at longer wavelengths, while those at shorter wavelengths shift much less and almost stop at specific wavelengths, meaning that the reflected light can be coupled to different modes. In the micrometer-diameter metal-dielectric-hybrid fiber tip, several modes are probably excited with close propagation constant because of the metal cladding [45]. Some modes are well confined in the tip and have negligible field overlap with the liquid while other modes are not. The different valleys and peaks correspond to the coupling between these different forward and backward propagating modes, with different response properties to any outer environment change [10].

The metal-dielectric-hybrid grating showed RI sensitive (a in Fig. 10(a)) and insensitive (d in Fig. 10(a)) behavior for different resonant modes [10]. S_a of the sensitive channel (125 nm/RIU) is one order of magnitude larger than that of a nonmetallic SCMG with the same radius whereas S_a of the insensitive channel (8 nm/RIU) is one order of magnitude smaller. This can be attributed to the fact that the introduction of metal film causes the MF to support both surface guided modes (which have a larger modal overlap with the ambient medium, a and b in Fig. 10) and bound modes (where most of the energy is located in the dielectric core, c and d in Fig. 10). The smallest sensitivity can be further decreased to nearly zero by

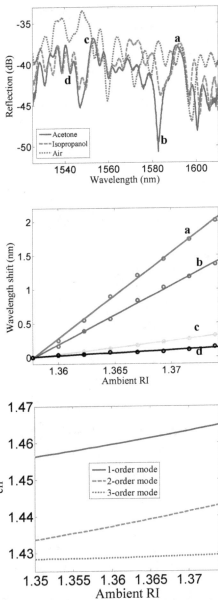

Figure 10. Top: measured reflection spectra of the metal-dielectric-hybrid grating when immersed in air, acetone, and isopropanol. a, b, c, d denote different peaks and valleys labelled. Centre: Dependence of wavelength shift on ambient RI for different modes in a metal-dielectric-hybrid SCMG. Bottom: Calculated effective index of one cladding mode and one core mode as a function of the outer liquid refractive index n_a. The radius of the fiber tip is assumed to be 3 μm with a golden coating 20 nm thick [10]. Copyright 2011 IEEE.

optimizing the tip grating profile and metal coating. Because of many different properties on the outer liquid refractive index, the metal-dielectric-hybrid grating can be applied as a multi-parameter sensor and the index-insensitive channel can be used to simultaneously measure temperature, pressure, and so on [10].

4.2. Temperature sensing

Although thermal post-processing and hydrogen loading have been shown to induce grating capable of standing temperatures as high as 1300 °C in conventional fibers [46], in MF thermometers, up to now, only SCMGs without polymer have been reported operating above 200 °C[9, 13]. The sensitivity of these components is around 20pm/°C, similar to the value predictable using Eq. (6). Figure 11 is the experimental characterization of the FIB milled SCMG demonstrated using the sample shown in Fig. 5(a). As the temperature increases, the Bragg wavelength red shifts. The extremely short SCMG length (~ 36.6µm) and wide operating range (~ 20 – 450°C) presents it as a promising candidate for detecting temperature change in ultra-small space.

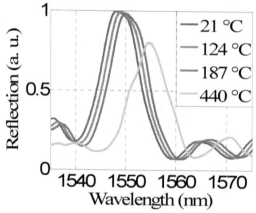

Figure 11. Reflection spectra of the FIB-milled SCMG in air at four different temperatures.

4.3. Strain/Force sensing

Although S_S remains almost the same for different MF diameters[26], S_F varies with the MF radius according to Eq. (9). A SCMG with diameter of 3.5 µm reaches a force sensitivity of ~ 1900 nm/N, which is more than three orders of magnitude compared to that of a conventional fibers [26]. If the Bragg wavelength can be detected to an accuracy of 0.05 nm, forces in the order of 10^{-5} N can be measured. For the sample reported in Fig. 5(d), where the silica constitute only a small fraction of the SCMG cross section, a further three orders of magnitude improvement in sensitivity is predicted, with S_F reaching values in excess of 10^6 nm/N, corresponding to forces of the order of nN. The SCMG strain/force sensors could offer attractive properties monitoring strain/force changes in power plant pipelines, airplane wings, and other civil engineering structures [19].

5. Conclusions

This review presented the fabrication, operating principles and applications of surface corrugated microfiber Bragg gratings (SCMGs). SCMGs can potentially outperform conventional FBGs because of their large evanescent field and compactness. Taking advantage of their extreme small size and unique geometry, SCMGs exhibit beneficial sensor properties such as fast response, high resolution and small-objection detections. SCMGs could find promising sensing applications in detecting parameter variations in ultra-small space [19].

Acronyms

FBG   fiber Bragg gratings
FIB   *focused ion beam*
fs   *femtosecond*
MF   *microfiber*
OSA   *optical spectrum analyzer*
SCMG   *surface-corrugated fiber grating*
SMF   *single mode fiber*
UV   *ultraviolet*

Author details

Fei Xu, Jun-Long Kou and Yan-Qing Lu
College of Engineering and Applied Sciences and National Laboratory of Solid State Microstructures, Nanjing University, Nanjing, P. R. China

Ming Ding and Gilberto Brambilla
Optoelectronics Research Centre, University of Southampton, Southampton, SO17 1BJ, United Kingdom

Acknowledgement

F. Xu and Y.-q Lu acknowledge the support from National 973 program under contract No. 2011CBA00200 and 2012CB921803, NSFC program No. 11074117 and 60977039, and the Priority Academic Program Development of Jiangsu Higher Education Institutions (PAPD). G. Brambilla gratefully acknowledges the Royal Society (London, U.K.) for his University Research Fellowship.

6. References

[1]   K. O. Hill, B. Malo, F. Bilodeau, D. C. Johnson, and J. Albert, "Bragg gratings fabricated in monomode photosensitive optical fiber by UV exposure through a phase mask " Appl. Phys. Lett. 62, 1035-1037 (1993).

[2] S. J. Mihailov, C. W. Smelser, P. Lu, R. B. Walker, D. Grobnic, H. M. Ding, G. Henderson, and J. Unruh, "Fiber Bragg gratings made with a phase mask and 800-nm femtosecond radiation," Opt. Lett. 28, 995-997 (2003).

[3] S. A. Slattery, D. N. Nikogosyan, and G. Brambilla, "Fiber Bragg grating inscription by high-intensity femtosecond UV laser light: comparison with other existing methods of fabrication," J. Opt. Soc. Am. B-Opt. Phys. 22, 354-361 (2005).

[4] C. Y. Lin and L. A. Wang, "A wavelength- and loss-tunable band-rejection filter based on corrugated long-period fiber grating," IEEE Photon. Technol. Lett. 13, 332-334 (2001).

[5] G. Brambilla, F. Xu, P. Horak, Y. Jung, F. Koizumi, N. P. Sessions, E. Koukharenko, X. Feng, G. S. Murugan, J. S. Wilkinson, and D. J. Richardson, "Optical fiber nanowires and microwires: fabrication and applications," Adv. Opt. Photon. 1, 107-161 (2009).

[6] T. L. Lowder, K. H. Smith, B. L. Ipson, A. R. Hawkins, R. H. Selfridge, and S. M. Schultz, "High-Temperature Sensing Using Surface Relief Fiber Bragg Gratings," IEEE Photonic. Technol. Lett. 17, 1926-1928 (2005).

[7] D. Grobnic, S. J. Mihailov, D. Huimin, and C. W. Smelser, "Bragg grating evanescent field sensor made in biconical tapered fiber with femtosecond IR radiation," IEEE Photon. Technol. Lett. 18, 160-162 (2006).

[8] X. Fang, C. R. Liao, and D. N. Wang, "Femtosecond laser fabricated fiber Bragg grating in microfiber for refractive index sensing," Opt. Lett. 35, 1007-1009 (2010).

[9] J.-l. Kou, S.-j. Qiu, F. Xu, and Y.-q. Lu, "Demonstration of a compact temperature sensor based on first-order Bragg grating in a tapered fiber probe," Opt. Express 19, 18452-18457 (2011).

[10] J.-l. Kou, S.-j. Qiu, F. Xu, Y.-q. Lu, Y. Yuan, and G. Zhao, "Miniaturized metal-dielectric-hybrid fiber tip grating for refractive index sensing," IEEE Photon. Technol. Lett. 23, 1712-1714 (2011).

[11] M. Ding, P. Wang, T. Lee, and G. Brambilla, "A microfiber cavity with minimal-volume confinement," Appl. Phys. Lett. 99, 051105 (2011).

[12] M. Ding, M. N. Zervas, and G. Brambilla, "A compact broadband microfiber Bragg grating," Opt. Express 19, 15621-15626 (2011).

[13] J. Feng, M. Ding, J.-l. Kou, F. Xu, and Y.-q. Lu, "An optical fiber tip micrograting thermometer," IEEE Photon. J. 3, 810-814 (2011).

[14] Y. Liu, C. Meng, A. P. Zhang, Y. Xiao, H. Yu, and L. Tong, "Compact microfiber Bragg gratings with high-index contrast," Opt. Lett. 36, 3115-3117 (2011).

[15] K. P. Nayak, F. Le Kien, Y. Kawai, K. Hakuta, K. Nakajima, H. T. Miyazaki, and Y. Sugimoto, "Cavity formation on an optical nanofiber using focused ion beam milling technique," Opt. Express 19, 14040-14050 (2011).

[16] F. L. Kien, K. P. Nayak, and K. Hakuta, "Nanofibers with Bragg gratings from equidistant holes," J. Mod. Opt. 59, 274-286 (2012).

[17] K. Handa, S. Peng, and T. Tamir, "Improved perturbation analysis of dielectric gratings," Applied Physics A: Materials Science & Processing 5, 325-328 (1975).

[18] T. Erdogan, "Fiber grating spectra," IEEE J. Lightwave Technol. 15, 1277-1294 (1997).

[19] J.-L. Kou, M. Ding, J. Feng, Y.-Q. Lu, F. Xu, and G. Brambilla, "Microfiber-Based Bragg Gratings for Sensing Applications: A Review," Sensors 12, 8861-8876 (2012).

[20] F. Xu, P. Horak, and G. Brambilla, "Optimized Design of Microcoil Resonators," Journal of Lightwave Technology 25, 1561-1567 (2007).

[21] F. Xu, P. Horak, and G. Brambilla, "Optical microfiber coil resonator refractometric sensor: erratum," Opt. Express 15, 9385-9385 (2007).

[22] I. M. White, H. Zhu, J. D. Suter, N. M. Hanumegowda, H. Oveys, M. Zourob, and X. Fan, "Refractometric sensors for lab-on-a-chip based on optical. ring resonators," Ieee Sensors Journal 7, 28-35 (2007).

[23] M. Adams, G. A. DeRose, M. Loncar, and A. Scherer, "Lithographically fabricated optical cavities for refractive index sensing," Journal of Vacuum Science & Technology B 23, 3168-3173 (2005).

[24] J.-l. Kou, J. Feng, L. Ye, F. Xu, and Y.-q. Lu, "Miniaturized fiber taper reflective interferometer for high temperature measurement," Opt. Express 18, 14245-14250 (2011).

[25] L. Sang-Mae, S. S. Saini, and J. Myung-Yung, "Simultaneous measurement of refractive Index, temperature, and strain using etched-core fiber Bragg grating sensors," IEEE Photon. Technol. Lett. 22, 1431-1433 (2010).

[26] T. Wieduwilt, S. Bruckner, and H. Bartelt, "High force measurement sensitivity with fiber Bragg gratings fabricated in uniform-waist fiber tapers," Meas. Sci. Technol. 22, 075201 (2011).

[27] W. Liang, Y. Huang, Y. Xu, R. K. Lee, and A. Yariv, "Highly sensitive fiber Bragg grating refractive index sensors," Appl. Phys. Lett. 86, 151122 (2005).

[28] A. Iadicicco, S. Campopiano, A. Cutolo, M. Giordano, and A. Cusano, "Refractive index sensor based on microstructured fiber Bragg grating," IEEE Photon. Technol. Lett. 17, 1250-1252 (2005).

[29] A. Iadicicco, A. Cusano, A. Cutolo, R. Bernini, and M. Giordano, "Thinned fiber Bragg gratings as high sensitivity refractive index sensor," IEEE Photon. Technol. Lett. 16, 1149-1151 (2004).

[30] R. Yang, J. Long, T. Yan-Nan, S. Li-Peng, L. Jie, and G. Bai-Ou, "High-efficiency ultraviolet inscription of Bragg gratings in microfibers," IEEE Photon. J. 4, 181-186 (2012).

[31] Y. Ran, Y.-N. Tan, L.-P. Sun, S. Gao, J. Li, L. Jin, and B.-O. Guan, "193nm excimer laser inscribed Bragg gratings in microfibers for refractive index sensing," Opt. Express 19, 18577-18583 (2011).

[32] F. Bilodeau, B. Malo, J. Albert, D. C. Johnson, K. O. Hill, Y. Hibino, M. Abe, and M. Kawachi, "Photosensitization of optical fiber and silica-on-silicon/silica waveguides," Opt. Lett. 18, 953-955 (1993).

[33] Y. Zhang, B. Lin, S. C. Tjin, H. Zhang, G. Wang, P. Shum, and X. Zhang, "Refractive index sensing based on higher-order mode reflection of a microfiber Bragg grating," Opt. Express 18, 26345-26350 (2010).

[34] F. Xu, G. Brambilla, and Y. Lu, "A microfluidic refractometric sensor based on gratings in optical fibre microwires," Opt. Express 17, 20866-20871 (2009).

[35] W. Ding, S. R. Andrews, T. A. Birks, and S. A. Maier, "Modal coupling in fiber tapers decorated with metallic surface gratings," Opt. Lett. 31, 2556-2558 (2006).

[36] M. C. Phan Huy, G. Laffont, V. Dewynter, P. Ferdinand, P. Roy, J.-L. Auguste, D. Pagnoux, W. Blanc, and B. Dussardier, "Three-hole microstructured optical fiber for efficient fiber Bragg grating refractometer," Opt. Lett. 32, 2390-2392 (2007).

[37] F. Xu, G. Brambilla, J. Feng, and Y.-Q. Lu, "A microfiber Bragg grating based on a microstructured rod: a proposal," IEEE Photon. Technol. Lett. 22, 218-220 (2010).

[38] R. R. Gattass and E. Mazur, "Femtosecond laser micromachining in transparent materials," Nat. Photon. 2, 219-225 (2008).

[39] J. C. Knight, "Photonic crystal fibres," Natur 424, 847-851 (2003).

[40] M. Bayindir, F. Sorin, A. F. Abouraddy, J. Viens, S. D. Hart, J. D. Joannopoulos, and Y. Fink, "Metal-insulator-semiconductor optoelectronic fibres," Natur 431, 826-829 (2004).

[41] G. Brambilla, F. Xu, and X. Feng, "Fabrication of optical fibre nanowires and their optical and mechanical characterisation," Electron. Lett. 42, 517-519 (2006).

[42] W. Streifer and A. Hardy, "Analysis of two-dimensional waveguides with misaligned or curved gratings," Quantum Electronics, IEEE Journal of 14, 935-943 (1978).

[43] F. Xu and G. Brambilla, "Preservation of Micro-Optical Fibers by Embedding," Jpn. J. Apl. Phys. 47, 6675-6677 (2008).

[44] J.-l. Kou, Z.-d. Huang, G. Zhu, F. Xu, and Y.-q. Lu, "Wave guiding properties and sensitivity of D-shaped optical fiber microwire devices," Applied Physics B 102, 615-619 (2011).

[45] J.-L. Kou, S.-J. Qiu, F. Xu, Y.-Q. Lu, Y. Yuan, and G. Zhao, "Miniaturized Metal-Dielectric-Hybrid Fiber Tip Grating for Refractive Index Sensing," IEEE Photon. Technol. Lett. 23, 1712-1714 (2011).

[46] J. Canning, M. Stevenson, S. Bandyopadhyay, and K. Cook, "Extreme Silica Optical Fibre Gratings," Sensors 8, 6448-6452 (2008).

Chapter 4

Long-Period Fiber Gratings in Active Fibers

David Krčmařík, Mykola Kulishov and Radan Slavík

Additional information is available at the end of the chapter

1. Introduction

Traditionally, long period fiber gratings (LPG) are made in passive optical fibers that have negligible loss. However, loss or gain that can be controlled via optical pumping adds a new degree of freedom and – as will be shown in this chapter – brings many new and interesting properties.

From the historical perspective, the first attempt to combine the fiber gain and LPG filtering characteristics was for gain flattening of an Erbium-doped fiber (EDF) amplifier by inscribing LPG directly into the active fiber [1]. At the same time, theoretical studies [2,3] showed that a proper level of loss/gain in the fiber core or cladding can modify the LPG transmission characteristics. Significant theoretical and experimental body of work has been published since with new emerging applications appearing.

In this chapter, we investigate the new phenomena brought by the presence of the loss/gain [2,3]. Following this, we look on practical possibilities how to obtain required gain in active optical fibers and show how to analyze such structures, in which (incoherent) noise from an amplifying fiber is simultaneously generated and diffracted at the LPG [4]. Finally, we discuss possible application of the LPG in active fibers.

2. Theoretical analysis

First, we analyze LPG using standard coupled mode theory in which we consider that the refractive index is a complex number in which the imaginary part represents the gain/loss. Following this analysis, we show approaches into rigorous modeling of the active gain medium that contains an LPG, including the spontaneous emission and amplified spontaneous emission generation, propagation, and diffraction.

2.1. Coupled mode equations considering gain/loss

Similarly to an LPG in a gain/loss-less fiber, LPG assists coupling between the core and a cladding mode at the wavelength for which the phase matching condition is satisfied for real part of the mode propagation constants:

$$\mathrm{Re}\left(\beta^{cl}\right) = \mathrm{Re}(\beta^{co}) - \frac{2\pi}{\Lambda}, \tag{1}$$

where $\beta^{cl} = (2\pi/\lambda)n^{cl}_{eff}+j\alpha^{cl}$ and $\beta^{co} = (2\pi/\lambda)n^{co}_{eff} +j\alpha^{co}$ are the propagation constants of the interacting cladding and core modes, respectively, λ is the wavelength, Λ is the grating period, n^{cl}_{eff} and n^{co}_{eff} are the effective refractive indices of the interacting cladding mode and the core mode, α^{cl} and α^{co} are the absorption (when positive) or amplification (when negative) factors of the cladding and the core modes, respectively. Propagation constants for the core and cladding modes can be straightforwardly calculated, especially when considering a step-index fiber refractive index profile [4-6]. The strength of the coupling between the two modes depends on the coupling coefficient calculated as a confinement factor between the field of the two interacting modes [7]:

$$\kappa(z) = 0.5\omega\varepsilon_0 n^2 \sigma(z) \int_0^{2\pi} d\phi \int_0^{r_1} [r(E_r^{cl} E_r^{co*} + E_\phi^{cl} E_\phi^{co*})]dr \tag{2}$$

where ω is the angular frequency, ε_0 is the permittivity of vacuum, $\sigma(z)$ is the LPG index modulation amplitude, E_r and E_ϕ are the radial and azimuthal parts of electric field. The asterisk signifies complex conjugate value. For non-uniform LPGs the index modulation amplitide generally varies along the fiber, so it depends on the distance z from the LPG input. In the following text we consider only LPGs which are uniform ($\sigma(z)=const$ over the entire LPG length).

The mode coupling process along the LPG is described by the following coupled mode equations [7]:

$$\frac{dA}{dz} = i\kappa B \exp(-2i\delta z), \tag{3}$$

$$\frac{dB}{dz} = i\kappa A \exp(2i\delta z), \tag{4}$$

$$\delta = \frac{\beta^{co} - \beta^{cl}}{2} - \frac{2\pi}{\Lambda}, \tag{5}$$

where $A(z)$ and $B(z)$ are slowly varying envelopes of the core mode and the interacting cladding mode, and z is the coordinate originated at the LPG input. The solution of these equations for a uniform LPG of length L and for the initial condition of input signal launched in the core mode only ($A(0) = 1$ and $B(z) = 0$) can be written in the form of the amplitude core-to-core mode transmission [2]:

$$t = \frac{A(L)}{A(0)} = \left[\cos(\gamma L) + j\frac{\delta}{\gamma}\sin(\gamma L)\right]\exp\left[jL\left(\beta^{co} - \delta\right)\right], \tag{6}$$

where γ is:

$$\gamma = \sqrt{\delta^2 + |\kappa|^2}. \tag{7}$$

2.2. Condition for critical coupling

In a standard fiber like SMF-28, propagation losses are low and can be completely neglected over few centimeters, which is a typical LPG length, even for high order cladding modes (provided the protective high-refractive index fiber jacket is stripped off). Therefore, the long period index perturbation along the fiber length provides constant optical power exchange between the coupled modes. An optical signal launched into the core at the LPG resonance wavelength will be completely coupled by the LPG into the cladding mode and then it will be completely coupled back into the core. This process will be repeated again and again as long as long-period perturbation is induced along the fiber. At the LPG resonance wavelength where the mismatch factor $\delta = 0$, the core-to-core mode transmission is

$$t = \cos(\kappa L). \tag{8}$$

The only way to provide full signal out-coupling from the core into the cladding mode is to terminate the LPG at discrete length values of $L = m\pi/(2\kappa)$, for m = 1, 2, 3, …

As we will describe now, the performance changes completely when there is an appropriate level of loss or gain in the fiber core or cladding. For example, for the case of cladding mode loss, optical signal is getting lost there and less optical power returns back into the core. However, the situation is more complicated/interesting as shown below.

At the wavelength where the mode matching condition (Eq.(1)) is satisfied, the mismatch factor becomes purely imaginary: $\delta=-j(\alpha_{cl} - \alpha_{co})/2$. Substituting this expression for mismatch factor at the resonance wavelength into Eq.(6), and equating it to zero, we get a relationship between the grating strength, κL, and the dimensionless loss factor, $q= (\alpha_{cl} - \alpha_{co})/(2\kappa)$:

$$t = \cos(\sqrt{1-q^2}\,\kappa L) + \frac{q}{\sqrt{1-q^2}}\sin(\sqrt{1-q^2}\,\kappa L) = 0 \tag{9}$$

The solution of Eq.(9) is presented (in dimensionless variables: κL and q) in Fig. 1. For each value of κL the curve gives us the optimum ratio between the attenuation and the coupling coefficient (characterized by q) that provides full signal light attenuation in the core mode. Thus, solution exists for any κL, unlike for a loss-less LPG, where full coupling is possible only at discrete values of κL. In other words, zero core mode transmission for an LPG of any strength κL can be found, depending on the loss/gain relations of the core and the interacting cladding modes.

It is worth analyzing which parts of the curve shown in Fig. 1 are practically attainable. For example, in a usual situation in which the core mode loss is smaller than the cladding mode loss, we get always $q>0$ and thus full coupling is possible (from Fig. 1) only for $\kappa L>\pi/2$. However, $q<0$ (and thus also full coupling at $\kappa L<\pi/2$, Fig. 1) can be, e.g., achieved by introducing loss into the core mode (that is larger than the cladding mode loss). Another interesting possibilities are considering core mode gain ($q>0$ and thus solutions limited to $\kappa L>\pi/2$) or cladding mode gain with the core mode unamplified ($q<0$ with solutions for $\kappa L<\pi/2$) [3]. The latter   could be practically achieved when considering double-clad fibers used for high-power fiber lasers, in which the cladding is doped by active ions [8].

Another interesting feature when considering gain/loss of core/cladding modes is that it may reduce interference sidelobes occurring outside of the resonance, obtaining a smooth, side-lobe free transmission spectrum [2].

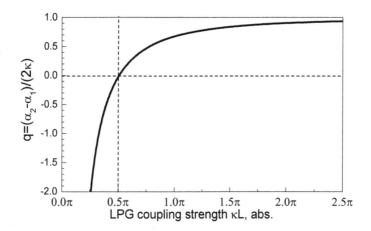

Figure 1. Required value of $q=(\alpha_{cl}-\alpha_{co})/(2\kappa))$ for obtaining full attenuation in the transmission spectrum of the core mode at the resonance wavelength (critical coupling) for a given LPG strength of κL.

To provide some physical insight into the mode coupling process in the presence of gain or loss, we show the grating transmission as a function of κL for various values of $q=\alpha_2/2\kappa$ (for clarity, only loss/gain in the cladding mode is considered here) – Fig. 2 [3]. The conventional sinusoidal transmission of loss/gain-free grating is shown as a solid curve. We can clearly see distinct types of transmission grating behavior depending on the magnitude of loss/gain.

2.3. Active fibers

The analysis of LPG in active fibers is based on a combination of two processes: coupled mode interaction (characterized by Eqs.(3-4)) and mode amplification (characterized by rate equations) [4]. To observe effects described in the previous section, high concentration of Er ions is required to achieve a significant gain over a length of an LPG that is usually relatively short (cm to tens of cm). Such high Er concentrations generally lead to cluster formations and up-conversion in the EDF, which has also to be taken into account.

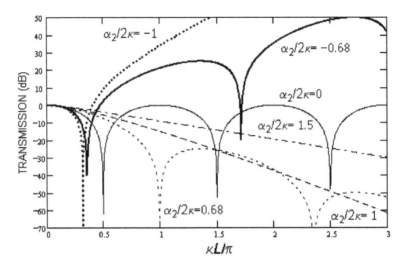

Figure 2. Variation of the grating transmission at the resonance wavelength with the grating strength κL for different $q=\alpha_2/2\kappa$ values.

There are several new phenomena occurring when considering not only the gain factor as in the previous part. The presence of gain/loss in the fiber with an LPG also changes resonance wavelength through modification of the effective refractive index of the coupled modes through Kramers-Kroning effect. Further, the presence of (amplified) spontaneous emission, (A)SE limits the maximum attainable transmission dip [4].

Here, the fibre is simulated by a two level system with the ground level and the meta-stable (excited) level. Due to high concentration of Er ions in the fibre it is necessary to take into account possible cluster formation and elevated occurrence of the up-conversion (process when the Er ions reach higher levels of excitation). Both cluster formation and up-conversion severely limits the maximum attainable gain for fibres with high concentration of Er ions and thus has to be considered.

In the up-conversion model we assume that the probability of the interaction of an ion with its neighbour is proportional to $N_2{}^2$, where N_2 is the number of Er ions in the excited state. Next model assumes cluster formation. In a cluster, only a single Er ion can contribute to the gain. In our analysis, clusters with more than two Er ions are not considered, as their contribution is generally negligible. In a combined model [9] the total number of ions contributing to the gain is:

$$N_2 = N_2^{SC} + N_2^P, \tag{10}$$

where $N_2{}^{SC}$ is an average number of single excited ions and $N_2{}^P$ is an average number of excited clusters. The average number of Er ions in not-excited state is $N_1 = N_{Er}-N_2$ (with N_{Er} total number of Er ions). Average number of single excited ions and excited clusters is:

$$N_2^{SC} = \frac{(1-2R)N_{Er}\sum_k \dfrac{P_k(z)\sigma_k^a \tau_{21}\Gamma_k}{hv_k A_{eff}}}{1 + C_{up}(1-2R)N_2^{SC}\tau_{21} + \sum_k \dfrac{P_k(z)}{P_k^{IS}}}, \tag{11}$$

$$N_2^P = \frac{2RN_{Er}\sum_k \dfrac{P_k(z)\sigma_k^a \tau_{21}\Gamma_k}{hv_k A_{eff}}}{1 + \sum_k \dfrac{P_k(z)}{P_k^{IS}} + \sum_k \dfrac{P_k(z)\sigma_k^a \tau_{21}\Gamma_k}{hv_k A_{eff}}}, \tag{12}$$

where

$$P_k^{IS} = \frac{hv_k A_{eff}}{(\sigma_k^a + \sigma_k^e)\tau_{21}\Gamma_k}. \tag{13}$$

In the above equations R is a percentage of ions in clusters, P_k is the peak power of signals and pump on v_k frequencies, τ_{21} is the metastable state lifetime, h is the Planck constant, A_{eff} is the effective core area and σ_k^a with σ_k^e are the absorption and emission cross-sections defined in the characteristics of the fiber. C_{up} [10] is the concentration independent and host dependent constant in m^3/s. Constant C_{up} can be determined by fitting the measured data. Γ_k is the overlap integral between the doped area and the optical mode field [11].

In an active medium we can observe a refractive index change due to the pump power. This translates into the slight wavelength shift of the coupling wavelength. The refractive index change is given by [12]:

$$\delta n = \left(\frac{\sigma_e' N_{Er}\alpha}{1+\alpha} - \frac{\sigma_a' N_{Er}}{1+\alpha} \right) \frac{\lambda\Gamma(\lambda)}{4\pi}, \tag{14}$$

$$\alpha = \frac{\sum_k P_k(z)}{P_{thr}}, \tag{15}$$

where P_{thr} is the threshold power for which the number of Er ions in the ground state is the same as in the meta-stable state ($N_1 = N_2$). P_k sums both the signal powers and the pump powers in forward and backward directions. Since the pump propagates mainly in the core, we have neglected any refractive index changes in the cladding. Real parts of the absorption coefficient σ_a' and the emission coefficient σ_e' are computed from σ_a and σ_e through Kramers-Kroning relations.

To include the active medium into the coupled-mode theory, it is necessary to define the signal absorption coefficient $g_a(z, v)$, signal emission coefficient $g_e(z, v)$ and pump absorption coefficient α_p [13]:

$$g_a(z,v) = \sigma^a(v)N_1(z)\Gamma(v) - BL, \tag{16}$$

$$g_e(z,v) = \sigma^e(v)N_2(z)\Gamma(v), \tag{17}$$

$$\alpha_p(z,v_p) = \sigma_p^a(v_p)N_1(z)\Gamma_p(v_p) + BL, \tag{18}$$

where the index p is attributed to the variables corresponding to the pump, and BL is the background loss. Here we assume a fiber with Er ions doped in the core only and thus that only the core mode is amplified since most of the power of the cladding modes propagates in the cladding. The core mode is amplified according to:

$$\frac{dP_A}{dz} = (g_e - g_a)P_A, \tag{19}$$

where P_A is the core mode intensity. The pump is then described similarly by:

$$\frac{dP_p^\pm}{dz} = -\alpha_p P_p^\pm. \tag{20}$$

The signs ± mean forward and backward pumping.

2.4. Amplified spontaneous emission and modes of pumping

As described above, more careful approach has to be taken for ASE. The coupled mode equations, Eqs.(3-4), are amplitude and phase dependent (describing coherent interaction), while the rate equations represented by Eqs.(16-20) are power intensity-dependent only (describing an incoherent interaction).

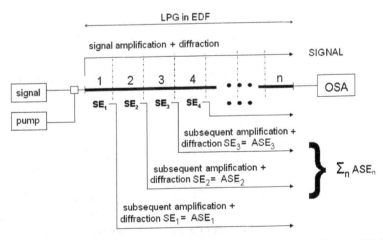

Figure 3. Description of SE and ASE treatment (SE – spontaneous emission, ASE – amplified spontaneous emission, OSA – optical spectrum analyzer).

ASE is seeded along the entire EDF length by spontaneous emission (SE) and subsequently amplified. In a standard EDF analysis, the fiber is divided into n segments first, the considered spectral bandwidth is divided into slots and each spectral component is propagated separately. Subsequently, the noise present at the input of the i-th segment, $i=1$, $2,..n$ (SE generated in the $(i-1)$-th segment and ASE amplified in the $(i-1)$-th segment) is amplified (forming ASE at the output of the i-th segment) and summed with SE generated in the i-th segment. Thus, at the output, the ASE is represented by a single number (for each spectral component). Here, we divide the fiber in the same manner as described above. However, we have to prevent summing of SE that was generated at different positions along the LPG: the SE generated at the beginning of the fiber undergoes diffraction along the entire length of the LPG, SE generated in its middle is diffracted at one half of the LPG, while the SE generated at its end is not diffracted at all. Thus, we take SE generated in the first segment and subsequently amplify it and diffract along the full length of the LPG. We continue similarly with the SE generated in the 2nd, 3rd ... nth segments. As a result, we have n contributions of the SE+ASE at the output (each coming from SE generated in the i-th segment, $i=1,2..n$ and subsequently amplified in the n-i segments) at the output. These contributions are incoherent (as each of them contains photons generated by SE) and thus can be summed in power intensity at the output, representing the ASE total power (for each spectral component). This approach is graphically shown in Fig. 3.

In practice, we generate SE and propagate ASE within each segment according to the equation:

$$\frac{dP_{ASE}^{\pm}}{dz} = \pm 2h\nu(\Delta\nu)g_e \pm (g_e - g_a)P_{ASE}^{\pm}, \tag{21}$$

where $\Delta\nu$ is the frequency slot. At the end of the segment the resulting ASE gets diffracted and new SE starts to arise at the beginning of the next segment.

Finally, we implemented also analysis for the contra-directional pumping configuration. In this scheme, the pump and the signal are propagated in opposite directions. For this configuration, the simulated data were obtained iteratively [4].

An example of the results using the theoretical analysis (populations of the meta-stable levels N_2 and ground levels N_1 of the EDF together with the pump power) is shown in Fig. 4. Here, we used parameters of the EDF later used in experiments – it is Liekki Er80-8/125 – more details can be found in Section 3.

Detailed analysis presented in [4] Fig. 5) reveals that optimal pumping scheme has to be chosen according to a specific application. Backward pumping is better when we want to limit the nonlinearities caused by an excessive power at the beginning of the fiber. On the other hand forward pumping shows better results for ASE suppression. For parameters considered, the LPG was always pumped relatively uniformly along its length. Therefore the transmission at the resonance wavelength is expected to be limited by the ASE formation rather than the non-uniform LPG strength that would be due to non-uniform pumping along the grating.

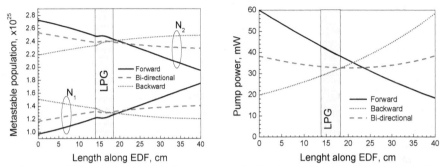

Figure 4. N_1 and N_2 populations (left) and pump power (right) along 40-cm long EDF sample with a 5-cm long LPG inscribed [position shown in the graphs] for total pump power of 60 mW and the input signal power of 100 µW.

3. Choice of EDF and fabrication techniques

Most of commercially-available EDFs have relatively high numerical aperture (0.2-0.4), which helps in obtaining good performance of these fibers. This, however, requires LPGs with sub-100-µm period, as the effective refractive index of the core mode is significantly higher than in low-numerical aperture fibers (e.g., telecom SMF-28 with numerical aperture of 0.12). Unfortunately, this is prohibitive for CO_2 [14] and arc [15] LPG inscription techniques – both of them operate on a principle of a local heating of the fiber and it is difficult to heat a fiber with 125 µm diameter over length significantly shorter than that. The techniques suitable are UV-writing [16] and fs-laser writing [17]. However high-gain EDF are doped with Al or P rather than Ge that is normally responsible for high UV photosensitivity. Thus, fs writing [17] seems as the most suitable method. However, to the best of our knowledge, this technique has not been used yet to fabricate LPG in EDFs; e.g., in [17], fs-writing was used for fiber Bragg grating fabrication.

In order to observe effects described in our theoretical analysis, e.g., tuning of the mode coupling interaction strength, it is necessary to provide enough gain across the LPG length that is typically 3-20 cm long. Practically, values of at least several dB per the LPG length are required, which is somehow more than in most commercially-available EDFs.

Fortunately, there are commercially available fibers that have low numerical aperture allowing use of relatively low cost techniques such as CO_2 and arc that do not require UV-photosensitivity and have high gain per unit length (tens of dB per meter). An example is Er80-8/125 from Liekki, Finland, that has 1532 nm peak absorption of 80 dB/m and numerical aperture of 0.13.

4. Experimental analysis

In this subsection we show first how gain alters the transmission characteristics of an LPG in line with predictions made in Section 2.2. Following that, we compare experimental results and those obtained using the rigorous theoretical simulations discussed in Section 2.3.

4.1. Tuning of an LPG via optical gain control

This feature was demonstrated in [18] in which a 13.5 cm long LPG of 480 μm period was inscribed into a 15-cm piece of EDF (Er80-8/125 from Liekki, Finland). Fig. 5 shows obtained results for an LPG with kL=0.52π. The theoretical predictions with the gain factor being 'fitted' to get good agreement with the experiment are also shown in Fig. 5. As we see, by controlling the gain we can obtain full coupling (at the pump power of 32 mW), although the grating has strength of kL=0.52π, exactly as predicted in the theoretical analysis shown above. It is worth mentioning that as the pump current is increased from 0 mW to 32 mW, the off-resonance transmission is increased while the resonance transmission decreases. Indeed, this effect cannot be observed by simply cascading a passive LPG with an EDF-based amplifier.

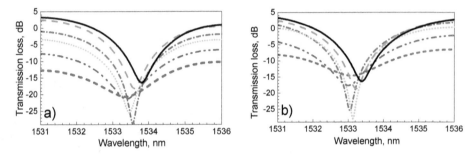

Figure 5. Measured (a) and calculated (b) transmission of LPG at kL=0.52π for input signal of 0.5 mW and different pump power levels (0 mW magenta, 8 mW blue, 17 mW green, 32 mW red, 145 mW orange, 220 mW black).

In practice, LPG-induced dip depth was varied from 7 dB (under-critical coupling) to the maximum of -28 dB (critical coupling) back to -18 dB (over-critical coupling) by varying the pump power between 0 and 220 mW.

4.2. LPG performance due to ASE and an EDF: comparison with rigorous analysis

Here we show experimental results considering an EDF theoretically modeled in the Section 2.3. Our LPG structure is shown in Fig. 6. The chosen LPG period of 480 μm corresponds to coupling into the 7th cladding mode.

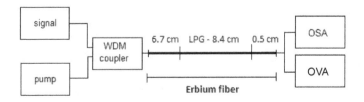

Figure 6. Experimental setup. OSA is optical spectrum analyzer; OVA is optical vector analyzer.

The theoretical and experimental results are shown in Fig. 7. For the critical coupling, there is about -33 dB of signal measured at the resonance in the core mode at the output, which is in a reasonable agreement with the theory that predicts -36 dB, Fig. 7b. These values are, however, significantly higher that those measured in a passive LPG, where <-60 dB was observed [14].

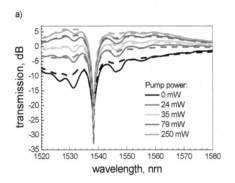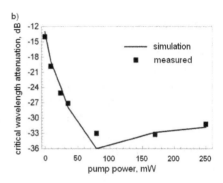

Figure 7. Left: experimental (solid) and theoretically computed (dashed) transmissions for various pump powers. Right: dependence of the maximum attainable resonant dip on the pump power (measured: dots; predicted: line)

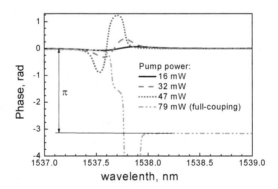

Figure 8. Phase characteristics of the fabricated LPG measured with optical vector analyzer (OVA, Fig. 6) for various pump powers.

To confirm that the lower level of the transmission loss for an active-fiber based LPG as compared to a passive-fiber-based LPG is due to ASE rather than any other effect (e.g., non-uniformity of the LPG due to pump power depletion across the LPG length), we measured the phase of the transmitted light around the resonant wavelength, Fig. 8. Theory predicts that full coupling should be manifested itself as a π phase jump across the resonance. As seen in Fig. 8, a π phase-shift was observed confirming the grating is operated at the critical coupling condition in which all the light from the fiber core mode is out-coupled. This confirms that the residual light in the core mode at the resonance wavelength (which is close to -33 dB, Fig. 7) has to be due to ASE rather than due to limited coupling ability of the LPG.

5. Applications

Several applications of LPGs in active fibers have been proposed. LPG directly-written into an EDF was suggested to perform flattening of the uneven EDF gain [1]. Tuning of the transmission characteristics of an LPG via optical pumping, which has been discussed in detail above, was reported in several reports (e.g.,[18-20]). This may be of interest in many applications, including all-optical signal processing, where besides active control of the grating parameters, it is advantageous to simultaneously filter and compensate for the loss that is due to the filtering process. We discuss in detail an example of an all-optical signal processing (all-optical differentiator) [20] later. Other applications are in fiber lasers, taking advantage of the fact that cladding modes have special dispersion properties that are suitable for controlling dispersion in mode-locked fiber lasers and also have large modal areas that can be exploited in high-power fiber lasers and amplifiers [21]. Another application is in using higher order modes for high-power optical amplification. In [16], the pump and the signal were simultaneously converted into a higher-order mode (a cladding mode of an inner-cladding of a dual-cladding fiber) – the LPGs for mode conversion were written in the EDF, which was doped with active ions simultaneously in the core and in the cladding.

5.1. All optical differentiation

Differentiation of the Nth order is mathematically described as $\partial^n u(t)/\partial t^n$, where $u(t)$ is the complex envelope of an arbitrary input signal spectrally centered at ω_0. The corresponding Fourier spectrum can be expressed as $(-j(\omega-\omega_0))^N U(\omega)$, where $U(\omega)$ is the spectrum of the input signal $u(t)$, ω is the optical frequency ($\omega-\omega_0$ is then the baseband). Therefore N-th order differentiation in optical domain can be obtained using a linear optical filter with a spectral transfer function proportional to $(-j(\omega-\omega_0))^N$. Unfortunately, such filter has zero transmission at the signal carrier ω_0, resulting in poor energetic efficiency (EE) defined as the ratio of the signal power after and before the differentiator. In practice, it was found that EE is <5% for an LPG-based differentiator [22]. This was confirmed by later experiments, in which some applications of this device for all-optical signal processing [23] were suggested. High order differentiation can be also realized using LPGs [24] by introducing π phase-shifts along the LPG. However, EE is getting worse as the order of the differentiation increases [24]. Obviously, LPG written in an active fiber could be very helpful in overcoming this issue.

Let us discuss advantages of LPGs directly written into active fibers on an example of optical differentiators. The use of a passive grating in conjunction with an amplifier at its output could lead to high levels of generated ASE inside the amplifier due to low EE of the differentiation process. In the case of pre-amplification of the signal prior to differentiation, emerging nonlinearities may severely impair the output signal. E.g., for pulses with FWHM < 1 ps, nonlinearities could be observed even for modest average powers. Ideal setup is thus a configuration where we perform amplification and filtration simultaneously which is the case of LPGs in active optical fibers. Indeed, the additional flexibility by tuning the LPG via controlling pump power should be also considered.

In general, the LPG structure in an active fiber is formed by a preamplifier (active fiber before the LPG), the active grating (LPG) and a post amplifier (active fiber after the LPG). Thus it is necessary to find optimal position of the grating within the active fiber to avoid the undesirable phenomena – low signal buried in the noise on one side and emerging nonlinearities on the other. In the process of designing it is necessary to take into account: total length of EDF, position of the grating within the EDF and input signal power. The total length of EDF influences the resulting gain of the output signal. Next two parameters (position of LPG and input signal power) are to be considered in combination for minimizing the ASE and avoid nonlinearities.

For theoretical analysis [4] we have chosen LPG sample with the following parameters [20]: coupling into the seventh cladding mode (the required grating period Λ=470 μm to obtain maximum coupling at 1540 nm); LPG length of 8 cm, which corresponds to the available bandwidth for differentiation to 500 GHz; input signal is Gaussian with FWHM=1.6 ps; length of EDF S=55 cm and pumping at 976 nm with power of 200 mW. In the simulation two levels of signal power were used – low (100 μW) and high (1 mW), because studied characteristics significantly differ for different levels of signal.

Energetic efficiency as a function of LPG position within EDF for two levels of signal power is shown in Fig. 9a.

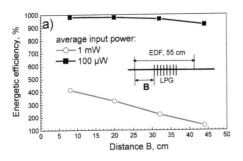
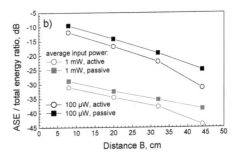

Figure 9. (a) Energetic efficiency for two levels of input signal power, (b) noise characteristics for EDF with embedded passive or active LPG.

For relatively low input powers the EE is practically independent from position of the grating along the fiber. For higher signal powers the dependence on the position of LPG is clearly visible (when the grating is at the beginning or end of the fiber, EE reaches 400%and 120%, respectively). The higher EE for the grating at the beginning of EDF is compensated by worse noise characteristics as shown on Fig. 9b.

For low input power (100 µW) the ratio of ASE to the total energy (ASE+signal) at the resonance drops from -12 dB (for LPG at the beginning of EDF) to -31 dB (for LPG at the end of EDF). For higher input signal power the ASE suppression is better ranging from -31 dB (LPG at the beginning of the EDF) to -44 dB (LPG at the end of the EDF). Thus, an LPG placed at the end of EDF gets the lowest ASE, while unfortunately also having the lowest EE. This trade-off between ASE suppression and EE has to be considered depending on the requirements of a particular application.

It is interesting to compare the studied design (LPG within EDF) with a design in which a passive LPG (written in a passive fiber) would be combined directly with an pre- and post-amplifier. We consider preamplifier of length B, passive LPG with the same length of 8 cm and post-amplifier with length S-B in order to get the same amplification at the end of the structure. The input signal power is the same for both cases (passive and active LPG). Comparison, Fig. 9b, shows that the passive model has – in contrast to the active model – 2 to 4 dB worse noise characteristics.

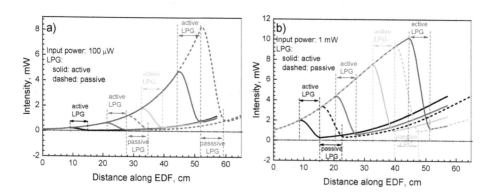

Figure 10. Power evolution for four different relative positions of the LPG along the EDF (black, red green and blue lines) for (a) 100 µW and (b) 1 mW input power. The performance considering a passive-fiber based LPG is shown as dashed lines.

For keeping nonlinearities low, it is necessary to minimize signal peak power along the EDF, which practically means keeping approximately constant level of the signal along the EDF. Fig. 10a shows the signal intensity for different position of the LPG within the EDF and the input signal power of 100 µW. At the same time it is shown that for the passive grating and the same length of EDF there are positions within the fiber with increased level of power and hence possible nonlinearities. Similarly we can observe such increased levels also for input signal power of 1 mW (Fig. 10b), although they are not as significant as in case of the weak input power. From Fig. 10 one can see that for preserving the same level of intensity along the fiber it is advantageous to place the LPG close to the middle or in the first quarter of the EDF for relatively low (100 μW) and high (1 mW) input signal power levels, respectively.

Experimental realization of a differentiator in an active LPG was reported in (Krcmarik et al., 2009). The differentiator was optimized for bandwidth of 1 THz (LPG length of 3.9 cm) and relative high input power levels (LPG within the first quarter of the EDF). The entire EDF length was 35 cm to obtain EE in excess of 100% (so-called 'loss-less differentiation'). Experimental results obtained using 0.9-ps FWHM pulses of 2 mW average power are shown in Fig. 11. For comparison, performance of an ideal differentiator is also shown there.

EE of 151% was measured. However, when the input signal was launched from the opposite side of the fiber, the EE dropped to 72%, exactly in line with the theoretical predictions discussed earlier. The processed pulse demonstrates very close fit to the theoretically predicted waveform. A sharp π-phase jump in the LPG transmission necessary for accurate differentiation was also observed.

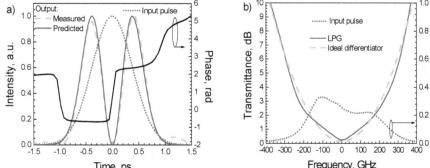

Figure 11. (a) Experimental and theoretical performance in time domain [20] measured phase characteristics [25] are also shown, (b) transmittance of the LPG in the vicinity of the resonance wavelength and input pulse spectrum showing that it fits entirely into the spectral region where the LPG transmittance follows the theoretically required shape.

For evaluation of the error between the ideal and the measured waveform we used square quadratic deviation formula [26]. For forward direction the error was 2.4% (Fig. 11) and for the reversed direction the error was 2.9%.

Realizing 'loss-less' higher-order differentiators, as well as characterizing experimentally the sensitivity to non-linearities are subject to further research.

6. Conclusions

Simultaneous diffraction and gain brings many interesting new features that are still waiting to be exploited. This chapter gives a brief overview over the state-of-the art and summarizes the key properties of long period gratings made in active optical fibers.

Author details

David Krčmařík
Institute of Photonics and Electronics AV CR, v.v.i, Praha, Czech Republic

Mykola Kulishov
HTA Photomask, 1605 Remuda Lane San Jose, CA, U.S.A.

Radan Slavík
Institute of Photonics and Electronics AV CR, v.v.i, Praha, Czech Republic
Optoelectronics Research Centre, University of Southampton, Southampton, Great Britain

7. References

[1]   Singh R, Sunanda, Sharma EK. Gain flattening by long period gratings in Erbium doped fibers. Optics Communications 2004; 240 (1-3), 123-132.

[2]   Daxhelet X, Kulishov M. Theory and practice of long-period gratings: when a loss becomes a gain. Optics Letters 2003; 28 (9) 686-688.

[3]   Kulishov M, Grubsky V, Schwartz J, Daxhelet X, Plant DV. Tunable waveguide transmission gratings based on active gain control. IEEE Journal of Quantum Electronics 2004; 40 (12), 1715-1724.

[4]   Krčmařík D, Slavík R, Karásek M, Kuslishov M. Theoretical and experimental analysis of long-period fiber gratings made directly into Er-doped active fibers. Journal of Lightwave Technology 2009; 27 (13), 2335-2342.

[5]   Marcuse D. (1991). *Theory of dielectric waveguides (2nd edition) – chapter 2*, Academic Press, ISBN 0124709516, New York

[6]   Kong M, Shi B. Field solution and characteristics of cladding modes of optical fibers. Fiber and Integrated Optics 2006; 25 (4), 305-321.

[7]   Erdogan T. Cladding-mode resonances in short- and long-period fiber grating filters. JOSA A 1997; 14 (8), 1760-1773.

[8]   Thyagarajan K, Anad JK. A novel design of an intrinsically gain-flattened erbium doped fiber. Optics Communications 2000; 183 (5-6), 407-413.

[9]   Jiang C, Hu W, Zeng Q. Numerical analysis of concentration quenching model of Er^{3+}-doped phosphate fiber amplifier. IEEE Journal of Quantum Electronics 2003; 39 (10), 1266-1271.

[10] Blixt P, Nilsson J, Carlnas T, Jaskorzynska B. Concentration-dependent upconversion in Er^{3+}-doped fiber amplifiers: experiments and modeling. IEEE Transaction Photonics Technology Letters 1991; 3 (11) 996-998.

[11] Myslinky P, Nguyen D, Chrostowski J. Effects of concentration on the performance of Erbium-doped fiber amplifiers. Journal of Lightwave Technology 1997; 15 (1), 112-120.

[12] Desurvire E. Study of the complex atomic susceptibility of Erbium-doped fiber amplifiers. Journal of Lightwave Technology 1990; 8(10), 1517-1527.

[13] Karásek M., Čtyroký J. Design considerations for Er^{3+}-doped planar optical amplifiers in silica on silicon. Journal of Optical Communications 1995; 16 (3), 115-118.

[14] Slavík R. Extremely deep long-period fiber grating made with CO_2 laser. IEEE Photonis Technology Letters 2006; 18 (16), 1705-1707.

[15] Rego G, Falate R, Santos JL, Salgado HM, Fabris JL, Semjonov SL, Dianov EM. Arc-induced long-period gratings in aluminosilicate glass fibers. Optics Letters 2005; 30 (16), 2065-2067.

[16] Nicholson JW, Fini JM, DeSantolo AM, Monberg E, DiMarcello F, Fleming J, Headley C, DiGiovanni DJ, Ghalmi S, Ramachandran S. A higher-order-mode Erbium-doped-fiber amplifier. Optics Express 2010; 18(17), 17651-17657.

[17] Wikszak E, Thomas J, Burghoff J, Ortac B, Limpert J, Nolte S, Fuchs U, Tünnermann A. Erbium fiber laser based on intracore femtosecond-written fiber Bragg grating. Optics Letters 2006; 31 (16), 2390-2392.

[18] Slavík R, Kulishov M. Active control of long-period fiber-grating-based filters made in erbium-doped optical fibers. Optics Letters 2007; 32 (7), 757-759.

[19] Quintela A, Quintela MA, Jauregui C, Lopez-Higuera JM. Optically tunable long-period fiber grating on an Er^{3+} fiber. IEEE Photonics Technology Letters 2007; 19 (10), 732-734.

[20] Krčmařík D, Slavík R, Park Y, Kulishov M, Azaña J. First-order loss-less differentiators using long period gratings made in Er-doped fibers. Optics Express 2009; 17 (2), 461-471.

[21] Sáez-Rodriguez D, Cruz JL, Díez A, Andrés MV. Fiber laser with combined feedback of core and cladding modes assisted by an intracavity long-period grating. Optics Letters 2011; 36 (10), 1839-1841.

[22] Kulishov, M., Azaña, J. (2005). Long-period fiber gratings as ultrafast optical differentiators. *Optics Lett.*, Vol. 30, No. 20, pp. 2700-2702, ISSN 0146-9592

[23] Slavík, R., Park, Y., Azaña, J. (2007). Tunable dispersion-tolerant picosecond flat-top waveform generation using an optical differentiator. *Opt. Express*, Vol. 15, No. 11, pp. 6717-6726, ISSN 1094-4087

[24] Kulishov M, Krčmařík D, Slavík R. Design of terahertz-bandwidth arbitrary-order temporal differentiators based on long-period fiber gratings. Optics Letters 2007; 32 (20), 1715-1724.

[25] Park Y, Li F, Azaña J. Characterization and optimization of optical pulse differentiation using spectral interferometry. IEEE Photonics Technology Letters 2006; 18 (17), 1798-1800.

[26] Ngo NQ, Yu SF, Tjin SC, Kam CH. A new theoretical basis of higher-derivative optical differentiators. Optics Communications 2003; 230 (1-3), 115-129.

Chapter 5

Optical Fibre Long-Period Gratings Functionalised with Nano-Assembled Thin Films: Approaches to Chemical Sensing

Sergiy Korposh, Stephen James, Ralph Tatam and Seung-Woo Lee

Additional information is available at the end of the chapter

1. Introduction

Optical techniques are considered as powerful tools for the development of chemical and biological sensors, covering a wide range of applications including bio-chemical and food analysis, and environmental and industrial monitoring [1, 2, 3]. Optical fibre sensors, as a result of their advantages such as high sensitivity, compactness, remote measurement and multiplexing capabilities, have attracted a great deal of attention for the development of refractometers and chemical sensors [4, 5, 6]. Refractometers and chemical sensors based on optical fibre gratings, both fibre Bragg gratings (FBGs) and long period gratings (LPGs), have been extensively employed for refractive index measurements and monitoring associate chemical processes since they offer wavelength-encoded information, which overcomes the referencing issues associated with intensity based approaches.

FBG based approaches have exploited side polished [7] and thinned [8,9] optical fibres to expose the evanescent field of the mode propagating in the core of the fibre to the surrounding medium, such that the Bragg wavelength becomes sensitive to the surrounding refractive index. The response of such sensors is non-linear, with a maximum predicted sensitivity of approximately 200 nm/refractive index unit (RIU) for indices close to that of the cladding of the optical fibre [8]. Experimental investigation revealed a sensitivity of 70 nm/RIU with a corresponding resolution of order 10^{-4} RIU, assuming that the Bragg wavelength is measured with a resolution of 1 pm [9].

The effective refractive indices of the modes of the cladding of the optical fibre are inherently sensitive to the surrounding refractive index. Tilted fibre Bragg gratings (TFBGs) and LPGs allow the controlled, resonant coupling of light from the core of the optical fibre into cladding modes. The characteristics of the resonant coupling are modified by changes in the surrounding refractive index, affording the ability to form refractive index sensors without altering the geometry of the optical fibre.

TFBGs facilitate the coupling of the propagating core mode to backward propagating cladding modes. As the coupling wavelength and efficiencies are dependent upon the properties of the cladding modes, the resonance features in the TFBG transmission spectrum exhibit sensitivity to surrounding refractive index [10,11]. Analysis of the transmission spectrum facilitates measurements with resolution of 10-4 RIU [11]. The use of thin film coatings of refractive index higher than that of the cladding on both polished FBGs and TFBGS has been shown to allow the RI range over which the devices show their highest sensitivity to the tuned to lower values [10,12,13].

LPGs promote coupling between the propagating core mode and co-propagating cladding modes, i.e. work as transmission gratings and are more attractive for practical applications as compared to FBGs, owing to lower cost of analytical instruments used to interrogate them. The high attenuation of the cladding modes results in the transmission spectrum of the fibre containing a series of attenuation bands centred at discrete wavelengths, each attenuation band corresponding to the coupling to a different cladding mode, as shown in Figure 1 [14].

The refractive index sensitivity of LPGs arises from the dependence of the phase matching condition upon the effective refractive index of the cladding modes that is governed by Equation 1 [14]:

$$\lambda_{(x)} = (n_{core} - n_{clad(x)})\Lambda \tag{1}$$

where $\lambda_{(x)}$ represents the wavelength at which the coupling occurs to the linear polarized (LP$_{0x}$) mode, n_{core} is the effective RI of the mode propagating in the core, $n_{clad(x)}$ is the effective RI of the LP$_{0x}$ cladding mode, and Λ is the period of the grating.

The effective indices of the cladding modes are dependent upon the difference between the refractive index of the cladding and that of the medium surrounding the cladding. The central wavelengths of the attenuation bands thus show a dependence upon the refractive index of the medium surrounding the cladding, with the highest sensitivity being shown for surrounding refractive indices close to that of the cladding of the optical fibre, provided that the cladding has the higher refractive index [15]. For surrounding refractive indices higher than that of the cladding, the centre wavelengths of the resonance bands show a considerably reduced sensitivity [15].

The refractive index sensitivity of an LPG is dependent upon the order of the cladding mode that is coupled to, allowing the tuning of the sensitivity by appropriate choice of grating period, with 427.72, 203.18, 53.45, and 32.10 nm/RIU being reported for LPGs fabricated in single mode fibre (SMF) 28 with period 159, 238, 400, and 556 µm, respectively [16]. A further consideration is the geometry and composition of the fibre, with the sensitivity being shown to differ for step index and W profile fibres and a progressive three layered fibre [17]. New fibre geometries, such as photonic crystal fibres [18,19] and photonic band gap fibres [20] have also been investigated for the measurement of the refractive index of a liquid that fills the air channels. LPGs in liquid filled solid-core photonic bandgap fibre have been shown to exhibit a sensitivity of 17,900 nm/RIU to changes in the RI of the liquid [21], but the requirement to fill the fibre with the liquid of interest may limit application as a refractometer.

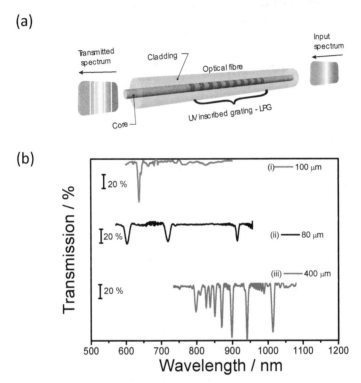

Figure 1. (a) Schematic illustration of the LPG structure and (b) transmission spectra of LPGs with different grating periods fabricated in an optical fibre of cut-off wavelength 670 nm (Fibrecore sm700): (i) 80 μm, (ii) 100 μm, and (iii) 400 μm [5].

In order to improve the sensitivity of the LPGs written in standard optical fibre configurations to surrounding RI, approaches such as tapering the fibre [22] or etching the cladding [23,24] have been investigated. Tapering the fibre to a diameter of 25 μm allowed the demonstration of an LPG with a sensitivity of 715 nm/RIU [22]. Etching the cladding of a section of standard single mode fibre containing an LPG from 125 μm to approximately 100 μm produced a sensitivity gain of 5 [25], while the etching an arc induced LPG to a diameter of 37 μm allowed the demonstration of a sensitivity of order 20,000 nm/RIU [26]. Approaches that require the processing of the fibre, such as polishing, etching and tapering, produce significant enhancements in sensitivity, but at the cost of requiring careful packaging to compensate for the inevitable reduction in the mechanical integrity of the device.

The deposition of thin film overlays, of thickness on the order of 200 nm, of materials of refractive index higher than that of the cladding has also been investigated for the enhancement of refractive index sensitivity [25,26]. It has been shown previously experimentally and theoretically [27,28] that the effective indices of the cladding modes, and thus the central wavelengths of the core-cladding mode coupling bands of LPGs, show a high sensitivity to the optical thickness of high refractive index coatings when the coating's optical thickness is such that one of the low order cladding modes is phase matched to a mode of the waveguide formed by the coating. This is termed the *mode transition region*, in which the cladding modes are reorganized, with each mode taking on the characteristics (effective index and electric field profile) of its adjacent lower order cladding mode [29]. The output from a numerical model of the influence of coating thickness on the effective indices of the cladding modes of an optical fibre is plotted in Figure 2, showing clearly the mode transition region.

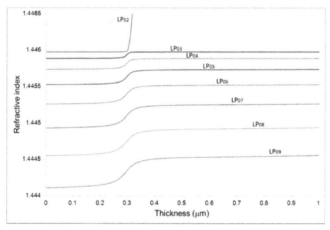

Figure 2. Plot of the effective refractive indices of the first 9 cladding modes as a function of the thickness of an overlay of refractive index. The model assumed values for the core and cladding refractive indices of 1.4496 and 1.446 respectively, a core radius of 3.5 μm, and that the refractive index of the overlay material was 1.57.

Biasing the optical thickness of the coating such that the LPG is operating in the *mode transition region* enhances the sensitivity of the cladding mode effective indices and thus the resonance wavelengths to the surrounding refractive index. A theoretical analysis explored the optimization of the refractive index sensitivity by selecting grating period, coating thickness and refractive index, predicting a sensitivity of 5980 nm/RIU [29].

In addition, the combination of optical fibers and nanomaterials provides a prospect for the fabrication of chemical sensors with high sensitivity and that offer specific response to targeted chemical species [30,31]. Achieving the coating thickness that provides optimized sensitivity requires control on the nm scale, which is why many reports have exploited the Langmuir Blodgett (LB) and layer-by-layer (LbL) electrostatic self assembly (ESA) coating deposition techniques, where a multilayer coating is deposited with each layer having a thickness of order 1 nm. Based on this principle, sensors for organic vapors, metal ions, humidity, organic solvents and biological materials have been reported [32, 33, 34]. Similarly to surface plasmon resonance (SPR) devices, LPG sensors can provide highly precise analytical information about adsorption and desorption processes associated with the RI and thickness of the sensing layer. For instance, the sensitivity of LPG sensors is in the same order of magnitude as SPR sensors, showing a sensitivity of ca. 1 nM for antigen–antibody reactions. An advantage of the LPG over SPR lies in the ability to fabricate a cheap and portable device that can be applied in various analytical situations.

The influence of the period of the LPG on the sensitivity of the mode coupling to perturbations such as changes in surrounding RI can be understood with reference to the phase matching condition, equation (1). Using the weakly guided approximation it is possible to determine the effective indices of the modes of the core and cladding, and, using equation 1, to generate a family of phase matching curves that describe the variation of the resonant wavelength with period of the LPG, an example is shown in Figure 3. It can be seen that the phase matching condition for each cladding mode contains a turning point. The resonance features in the LPG spectrum exhibit their highest sensitivity to external perturbations near the phase matching turning point. The response of the transmission spectrum of an LPG operating near the phase matching turning point external perturbation, in this case changes to the optical thickness of a coating deposited onto the optical fibre, are illustrated in Figure 4. In Figure 4a, the grating period is such that phase matching to the LP_{020} cladding mode is satisfied, but it is not possible to couple to the LP_{021} cladding mode, with a resulting LPG transmission spectrum of the form shown in Figure 4b. Changes in the optical thickness (product of the geometrical thickness and refractive index) of the coating cause an increase in the effective refractive index of the cladding modes, and the phase matching curves change accordingly, as illustrated in Figure 4c, resulting in the development of a resonance band corresponding to coupling to the LP_{021} cladding mode, and a small blue shift in the central wavelength of the LP_{020} resonance band, Figure 4d. Further increases in the optical thickness of the coating result in the further development of the LP_{021} resonance band, which subsequently splits into two bands, the so called dual resonance, Figures 4f and 4h. The small gradient of the phase matching curve at the phase matching turning point results in the LP_{021} resonance band exhibiting much higher sensitivity than the LP_{020} resonance band for this grating period. Thus the sensitivity of coated LPG sensors can be optimised by appropriate choice of the optical thickness of the coating and period of the grating to ensure that the LPG operates at both the mode transition region and the phase matching turning point.

Recently, LPG fibre sensors with porous coatings have attracted a lot of interest. For instance, a sol-gel film of SnO_2 of thickness of order 200 nm was deposited onto to an LPG, facilitating the demonstration of an ethanol gas sensor. The porosity and high RI of the coating material resulted in the LPG spectrum exhibiting a response to the diffusion of ethanol gas into the coating and the authors predicted that an optimized sensor would exhibit a detection limit of 100 ppb. Sol-gel coatings of SiO_2 and TiO_2 have been deposited onto LPGs, revealing a gain of 2 in the sensitivity to external RI, with the higher index TiO_2 coating offering the larger response (up to 1067.15 nm/RIU). The authors noted that the higher the index and thickness of the coating the more pronounced was the enhancement in LPG sensitivity compared to the equivalent uncoated LPG [16].

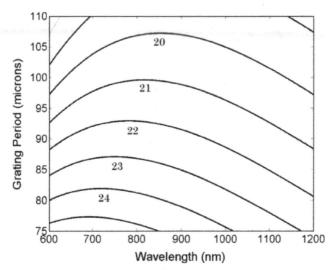

Figure 3. Phase matching curves for higher order cladding modes of an optical fibre of cut off wavelength 670 nm. The numbers refer to the order of the cladding mode.

Recently we have demonstrated a fibre optic refractive index sensor based on a long period grating (LPG) with a nano-assembled mesoporous coating of alternate layers of poly(allylamine hydrochloride) (PAH) and SiO_2 nanospheres [5, 6]. PAH/SiO_2 coatings of different thicknesses were deposited onto an LPG operating near its phase matching turning point in order to study the effect of the film thickness and porosity on sensor performance. The device showed a high sensitivity (1927 nm/RIU) to RI changes with a response time less than 2 sec over a wide RI range (1.3330–1.4906). The low refractive index of the mesoporous film, 1.20@633 nm, facilitates the measurement of external indices higher than that of the cladding, extending the range of operation of LPG based RI sensors. The ability of this device to monitor, in real time, RI changes during a dilution process was also demonstrated.

In this chapter we introduce a new approach to LPG based chemical sensing. A novel 2 stage approach to the development of the sensor is explored. The first stage involves the deposition of the mesoporous coating onto the LPG operating near the phase matching turning point. In the second stage a functional material, chosen to be sensitive to the analyte of interest, is infused into the base mesoporous coating. The mesoporous coating consists of a multilayer film of SiO_2 nanoparticles (SiO_2 NPs) deposited using the LbL technique. The initially low RI of the mesoporous coating, 1.2@633 nm, is increased significantly by the chemical infusion, resulting in a large change in the LPG's transmission spectrum. The sensing of ammonia in aqueous solution was chosen to demonstrate the sensing principle. The operation of the sensor was characterised using two functional materials, tetrakis-(4-sulfophenyl)porphine (TSPP) and polyacrylic acid (PAA). TSPP is a porphine compound that changes its optical properties (absorbance and RI) in response to exposure to ammonia, while PAA has been employed as a functional compound ammonia binding [35]. In the case of the PAA it is assumed that direct biding of ammonia to the COOH moiety will change RI of the mesoporous coating, while in the case of TSPP its desorption will result on RI change.

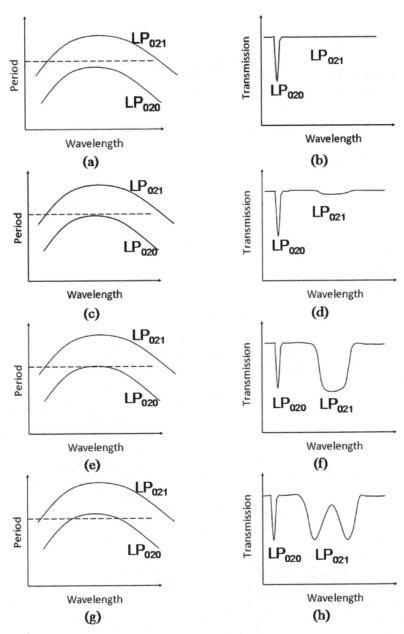

Figure 4. Optical features of the LP_{020} and LP_{021} cladding modes near the phase matching turning point. An increase in the effective index of the cladding modes, caused for example by an increase in the surrounding refractive index, or in the optical thickness of a coating deposited onto the fibre, cause the phase matching curves to move as shown, producing large changes in the coupling characteristics and transmission spectrum.

2. Sensor fabrication

Tetrakis-(4-sulfophenyl)porphine (TSPP, Mw = 934.99) was purchased from Tokyo Kasei, Japan. Poly(diallyldimethylammonium chloride) (PDDA, Mr = 200,000–350,000, 20% (w/w) in H2O; monomer Mw = 161.5 g mol−1), PAA$_{25}$ (Mw: 250,000, 35 wt% in H2O) and ammonia 30 wt% aqueous solution were purchased from Aldrich. A colloidal solution of silica nanoparticles (SiO$_2$ NPs), SNOWTEX 20L (40–50 nm), was purchased from Nissan Chemical. All chemicals were reagents of analytical grade, and used without further purification. Deionized pure water (18.3 MΩ·cm) was obtained by reverse osmosis followed by ion exchange and filtration in a Millipore-Q (Millipore, Direct-QTM).

A detailed description and reference to the optical properties of LPGs can be found elsewhere [5, 6, 36]. In this work, an LPG of length 30 mm with a period of 100 μm was fabricated in a single mode optical fibre (Fibercore SM750) with a cut-off wavelength of 670 nm using point-by-point UV writing process. The photosensitivity of the fibre was enhanced by pressurizing it in hydrogen for a period of 2 weeks at 150 bar at room temperature.

SiO$_2$ NPs were deposited onto the surface of the fibre using the LbL process, as illustrated in Figure 5a. As the LPG transmission spectrum is known to be sensitive to bending, for the film deposition process and ammonia detection experiments the optical fibre containing LPG was fixed within a special holder, as shown in Figure 5b, such that the section of the fibre containing the LPG was taut and straight throughout the experiments [36]. The detailed procedure of the deposition of the SiO$_2$ NPs onto the LPG and infusion of the TSPP compound has been previously reported [5]. Briefly, the section of the optical fibre containing LPG, with its surface treated such that it was terminated with OH groups, was alternately immersed into a 0.5 wt% solution containing a positively charged polymer, PDDA, and, after washing, into a 1 wt% solution containing the negatively charged SiO$_2$ NPs solution, each for 20 min. This process was repeated until the required coating thickness was achieved. When the required film thickness had been achieved (i.e. when the development of the second resonance band was observed with the fibre immersed into water), ca. after 10 deposition cycles, the coated fibre was immersed in a solution of TSPP or PAA as functional compound for 2 h, which was infused into the porous coating and provided the sensor with its specificity. Due to the electronegative sulfonic groups present in the TSPP compound, an electrostatic interaction occurs between TSPP and positively charged PDDA in the PDDA/SiO$_2$ film. On the other hand, PAA is usually considered as a promising sensor element for ammonia sensing, owing to the presence of free carboxylic function groups that lead to high affinity towards amine compounds [37]. After immersion into the TSPP and PAA solutions, the fibre was rinsed in distilled water, in order to remove physically adsorbed compounds, and dried by flushing with N$_2$ gas. The compounds remaining in the porous silica structure were bound to the surface of the polymer layer that coated each nanosphere. This effectively increased the available surface area for the compounds to bond to. The presence of functional chemical compounds increased the RI of the porous coating and resulted in a significant change in the LPG's transmission spectrum, consistent with previous observations for increasing the coating thickness [38]. All experiments have been conducted at 25°C and 50% of rH.

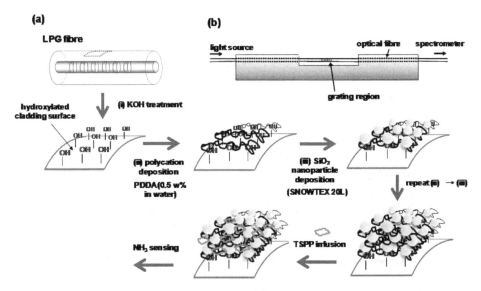

Figure 5. (a) Schematic illustration of the electrostatic self-assembly deposition process and (b) deposition cell with a fixed LPG fibre.

3. Mesoporous coating of long period gratings

Figures 6a and 6b show the surface morphology and cross-section of the 10-cycle $(PDDA/SiO_2)$ layer, referred to as $(PDDA/SiO_2)_{10}$, on a quartz substrate, respectively. As can be seen, the $(PDDA/SiO_2)_{10}$ film has a uniform surface consisting of SiO_2 NPs with an average diameter of 45 nm (Figure 6a). The film thicknesses obtained after 1 and 10 cycles of the $(PDDA/SiO_2)$ deposition process, determined from SEM cross-section measurements, are approximately 50 ± 2 nm and 450 ± 20 nm, respectively (Figures 6b and 6c). The pore size distribution of the $(PDDA/SiO_2)$ film indicates a well-developed mesoporous structure with a mean pore radius of 12.5 nm and specific surface area of 50 $m^2\,g^{-1}$ [5].

The transmission spectrum of the 100 μm period LPG undergoes changes due to the alternate deposition of SiO_2 NPs, as shown in Figure 7, which influences the effective RI of the cladding mode, as described previously [5]. When the LPG was in the silica colloidal solution, the resonance feature (at ca. 640 nm) corresponding to coupling to the LP_{020} cladding mode exhibited a small blue wavelength shift of 8.5 nm. As the optical thickness of the coating increased, it became possible to couple energy to the LP_{021} mode, with the corresponding development of the resonance band at ca. 800 nm [5], at the phase matching turning point for that mode. However, because of the low RI (1.20) [39] of the porous silica coating, this resonance feature is not well developed in water and is not present in air for this coating thickness.

When an uncoated LPG fibre with a period of 100 μm is immersed into water (RI=1.323), a blue shift of the LP_{020} resonance band of 3 nm is observed. However, the sensitivity of the

LP_{020} resonance band is much improved by coating the LPG with a $(PDDA/SiO_2)_{10}$ film, showing a blue shift of 7 nm when the LPG was immersed in water, along with the appearance of the well pronounced second resonance band, LP_{021} (Figure 8a). This is attributed to the increase in sensitivity of the LPG to the surrounding RI [16], being of interest when measuring the RI of low concentration aqueous solutions. The response to the RI changes is fast (< 2 s) and stable, as indicated by the measurement of the transmitted power at the centre of the LP_{021} resonance (Figure 8b).

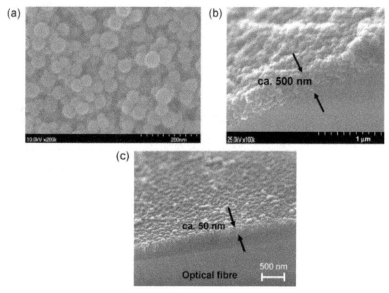

Figure 6. SEM images of the (a) surface morphology and (b) cross-section of the $(PDDA/SiO_2)_{10}$ film deposited on a quartz substrate before TSPP infusion; (c) cross section of a $(PDDA/SiO_2)_1$ film.

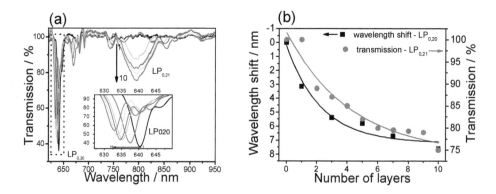

Figure 7. (a) Transmission spectra of the 100 μm period LPG in the colloidal SiO_2 solution in water after PDDA deposition (b) wavelength shifts and changes in transmission as a function of the number of deposition cycles for the LP_{020} and LP_{021} resonance bands, respectively; the curve is a guide to the eye only.

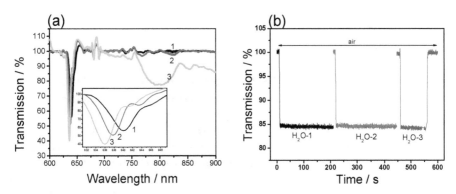

Figure 8. (a) Transmission spectra of the 100 μm period LPG under different conditions: black line, in air without coating; red line, in water without coating; green line, in water after deposition of the $(PDDA/SiO_2)_{10}$ film. (b) Dynamic changes of the transmission spectrum of the SiO_2 NP coated LPG measured at 800 nm in different phases from air to water.

The RI and film thickness of a 1-cycle $PDDA/SiO_2$ film, measured using ellipsometry, were 1.20 (at 633 nm) and 47±2 nm, respectively, which is in a good agreement with both the reported data for RI [39] and the thickness measured using SEM (see Figure 6b). It should be noted that dispersion of the RI in the wavelength range of 400–800 nm is negligible, with the RI value of 1.20±0.0001, and does not influence LPG sensor performance over the operational wavelength range of 600–900 nm. The deposition of the $PDDA/SiO_2$ layer was also monitored using the QCM technique and UV-vis spectroscopy (data not shown). A $(PDDA/SiO_2)_{10}$ film was deposited on two different QCM electrodes and a relative standard deviation of ±11% was obtained. The frequency linearly decreased as the number of the deposition cycles increased with average values of 1739±207 Hz and 30±10 Hz for the SiO_2 and PDDA layers, respectively. This corresponds to SiO_2 and PDDA masses of 1565 ng and 27 ng per each layer, respectively. The average thickness of the SiO_2 layer was 42±4 nm [36], which corresponds very well with the values determined using ellipsometry and SEM. A uniform $PDDA/SiO_2$ film was assembled on the quartz substrate, as revealed by the increase of the absorbance in the UV region with increment in the number of SiO_2 NP layers. The modulation observed in the absorption spectrum of the coated quartz slide was the result of the interference between light reflected from the front and rear surfaces of the coated slide, which introduces a channelled spectrum, the period and phase of which is dependent on the $PDDA/SiO_2$ film optical thickness. Thickness values of the $PDDA/SiO_2$ film deposited onto different substrates (quartz slab, silicon wafer and optical fibre) using a LbL method determined from SEM, QCM and ellipsometry measurements agreed well, within experimental error, regardless of the surface nature, indicating that the LbL method is an efficient tool for the deposition of uniform nano-thin films on different types of surfaces.

4. Employment of functional molecules and chemical sensing

The principle of operation of a coated LPG sensor operating at the phase matching turning point and applied to the detection of chemical components that can be bound through an electrostatic interaction with PDDA in the mesopores of the film has been discussed previously [5]. The resonance bands (LP$_{020}$ and LP$_{021}$) could be used for the detection of chemical components that can be bound through an electrostatic interaction with the cationic groups of PDDA in the mesopores of the film. When an LPG coated with a (PDDA/SiO$_2$)$_{10}$ film is immersed in a TSPP solution, the transmission spectrum undergoes significant changes. Figure 9a shows the transmission spectra recorded when the (PDDA/SiO$_2$)$_{10}$ film coated LPG was immersed in a 1 mM TSPP water solution. As the TSPP is infused into the film, the RI of the film increases (from 1.200 to ca. 1.540, measured using ellipsometry) and the phase matching condition for coupling to LP$_{021}$ is satisfied. A broad single attenuation band is developed rapidly (within 60 s), which subsequently splits in two bands as the RI of the coating increases in response to the TSPP infusion. The required time to complete the binding between the TSPP and PDDA moieties is less than 600 s (Figure 9b). The observed response indicates a large increase in the optical thickness of the film, which is a result of the increase in the RI of the film, as the TSPP is infused into the porous structure and adsorbed to the PDDA moiety between SiO$_2$ NPs.

The evolution of the transmission spectrum of the SiO$_2$ coated LPG when immersed in the TSPP solution is shown in the grey scale plot shown in Figure 10, where the transmission is represented by white and black, corresponding to 100% and 0%, respectively. The dark line at around 635 nm, which originates at a wavelength of 640 nm in the uncoated LPG, represents the resonance band that corresponds to the first order coupling to the LP$_{020}$ cladding mode. The discontinuity in the trace, at 60 s, occurred when the LPG was immersed in the solution and is a result of the LPG's sensitivity to the RI of the solution.

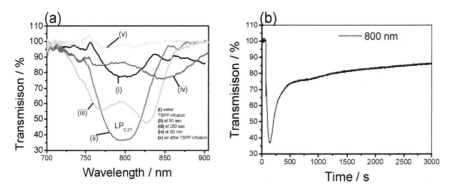

Figure 9. (a) Transmission spectra of the SiO$_2$ NP coated LPG and (b) the dynamic transmission change recorded at 800 nm when the SiO$_2$ NP coated LPG was immersed in a TSPP solution (1 mM in water).

In order to assess the sensitivity of the optical device, the (PDDA/SiO$_2$)$_{10}$ coated LPG was exposed to different concentrations of TSPP, and the results are shown in Figures 11a and 11b. The increase of the TSPP concentration from 10 to 1000 µM results in a decrease of the transmission measured at 800 nm, corresponding to the development of the LP$_{021}$ cladding mode resonance. This is also accompanied by a blue shift of the LP$_{020}$ resonance band, indicating the increase of the RI of the film. The response time of the sensor is observed to be slower at lower TSPP concentrations. For 1 mM TSPP concentration, the increase in transmission at 800 nm, shown in Figures 11b, is attributed to the splitting of the fully developed LP_{021} cladding mode resonance into dual resonance bands.

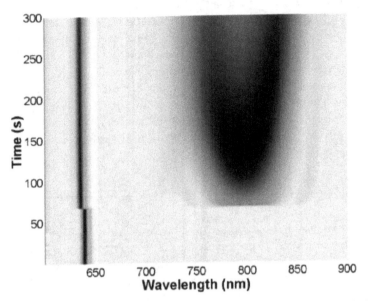

Figure 10. The evolution of the transmission spectrum of the SiO_2 NP coated LPG (period 100 μm), when immersed in an aqueous solution of TSPP (1 mM). The grey scale represents the measured transmission, with white corresponding to 100%, and black to 0%.

The ability to reuse the device was tested by removing the infused TSPP molecules from the film using an ammonia solution (ca. 1000 ppm). The spectrum was reverted to that observed for the $(PDDA/SiO_2)_{10}$ coated LPG before TSPP infusion. Subsequent immersion of the $(PDDA/SiO_2)_{10}$ coated LPG into the TSPP solutions allowed the results shown in Figure 11 to be reproduced.

A similar effect to that observed for the infusion of TSPP was observed when the $(PDDA/SiO_2)_{10}$ coated LPG was immersed into a PAA solution. The magnitude of the change, however, was smaller as compared to that induced by TSPP infusion, most plausibly being related to the lower RI of PAA (1.442) [40] as compared with that of TSPP (1.540) [5]. It should be noted that thickness of the $(PDDA/SiO_2)_{10}$ film was not changed on the infusion of the functional compounds, as revealed from SEM cross-sectional and ellipsometry measurements of the samples deposited onto the optical fibre before and after TSPP infusion (data not shown).

The infusion of the TSPP molecules into the $PDDA/SiO_2$ film deposited on a glass substrate was also investigated using UV-vis spectroscopy. Figure 12 shows the absorbance spectrum of the $PDDA/SiO_2$ film after infusion of the TSPP compound. As can be seen from Figure 12, the two Soret bands at wavelengths 419 and 482 nm, along with well-pronounced Q-band at 700 nm, are present, which indicates that the TSPP compound forms J-aggregates in the porous silica film [41,42]. It was previously confirmed that nano-assembled thin films with TSPP in J-aggregate state are particularly sensitive to ammonia gas [42]. In this work, mesoporous SiO₂ NPs films infused with TSPP were to be used to detect the presence of ammonia in aqueous solutions. In order to study the stability of the PDDA/SiO₂ films infused with TSPP in aqueous solutions, they were immersed into water several times, with the resultant changes in absorption spectra shown in Figure 12. The second Soret-band (482 nm) and Q-band (at 700 nm) disappeared when the film was immersed into water, accompanied by a decrease in the absorbance of the first Soret-band (419 nm), which indicates the partial removal of the adsorbed TSPP molecules from the film [42]. We can speculate that this phenomenon is a result of the cleavage of the J-aggregates of TSPP formed in the space between the SiO₂ NPs.

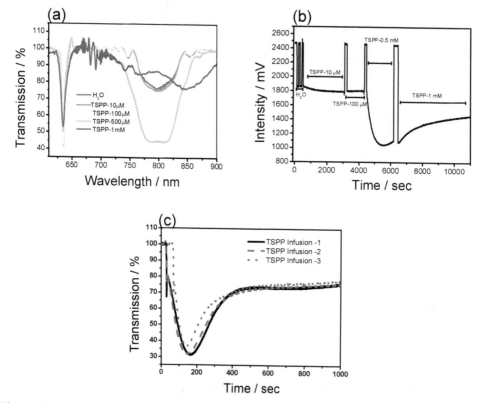

Figure 11. (a) Transmission spectra of the (PDDA/SiO₂)₁₀ coated LPG; (b) transmission change recorded at 800 nm in response to different concentrations of TSPP (from 10 μM to 1 mM in water) and (c) dynamic response to the three infusions of TSPP into the PDDA/SiO₂ porous film recorded at 800 nm; the infusion was conducted after complete removal of TSPP from the PDDA/SiO₂ using an NH₃ solution of concentration 1000 ppm.

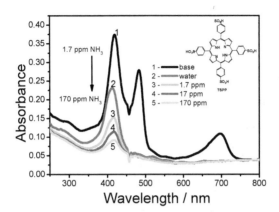

Figure 12. Absorption spectra of the (PDDA/SiO₂)₁₀ film infused with TSPP on a quartz plate: before immersion into H₂O (solid line) and after immersion into ammonia solutions of concentration from 0 ppm, 1.7 ppm, 17 ppm and 170 ppm (dashed lines); inset shows the structure of the TSPP.

The observed absorbance decrease reached a steady state after several immersions, indicating that only strongly bound TSPP molecules remained in the PDDA/SiO₂ porous film. Subsequently, immersion of the PDDA/SiO₂ film into aqueous solutions containing ammonia led to the further removal of TSPP from the film, resulting in a decrease of the absorbance at 419 nm in proportion to the ammonia concentration, as shown in Figure 12. The absorbance at 419 nm almost disappeared, at an ammonia concentration of 170 ppm.

The amount of TSPP infused inside the mesoporous film was estimated from QCM measurements to be 1135.6 ng (1.13 nmol). Consequently, considering the amount of adsorbed PDDA to be 250 ng (278 Hz for 10 cycles, 1.50 nmol as monomer unit of PDDA), the porous film contains sufficient binding sites for the infused TSPP molecules using the electrostatic interaction. When the QCM electrode that had been coated with a (PDDA/SiO₂)₁₀ film infused with TSPP was immersed into water and into ammonia, a similar trend to that measured using a UV-vis spectrophotometer was observed [36]. In particular, the frequency increased after each immersion into water (for 5 min) by up to a factor of 5, indicating the desorption of TSPP from the film; about 70% of the employed TSPP molecules were removed from the film [36]. Further immersion into water did not lead to a significant frequency change, suggesting that strongly bound TSPP (ca. 30% of the employed TSPP molecules) remained in the mesoporous film. When the film was exposed to ammonia solutions of different concentrations, further desorption of TSPP was observed and the total mass loss was 292 ng, indicating the almost complete removal of TSPP from the film [36]. Consequently, ca. 51 ng of TSPP remained in the film after ammonia treatment.

For ammonia detection, the sensing mechanism may be based upon changes in the RI of the coating resulting from chemically induced adsorption or desorption of the functional material to or from the mesoporous coating. Porphyrins have tetrapyrrole pigments [43] and their optical spectrum in the solid state is different to that in solution, due to the presence of strong π–π interactions [43]. Interactions with some chemical species can produce further spectral changes, thus creating the possibility that they can be used in the development of optical sensor systems. For instance, exposure of TSPP that has sulfonic functional groups to ammonia leads to the modification of the absorption spectrum [42].

5. Ammonia sensing

The sensitivity to ammonia in water of an LPG coated with a $(PDDA/SiO_2)_{10}$ film that was infused with TSPP was characterized by sequential immersion of the coated LPG into ammonia solutions with different concentrations (0.1, 1, 5 and 10 ppm). The lower ammonia concentrations were prepared by dilution of the stock solution of 28 wt%. In order to assess the stability of the base line, the coated LPG was immersed several times into 150 μL of pure water. The decrease of attenuation of the second resonance band, LP_{021}, at 800 nm, indicates the partial removal of the adsorbed TSPP molecules as discussed above. The equilibrium state was achieved after several exposures into water. For the ammonia detection, the LPG fibre was exposed into a 150 μL ammonia solution of 0.1 ppm, followed by drying and immersion into ammonia solutions of 1, 5 and 10 ppm.

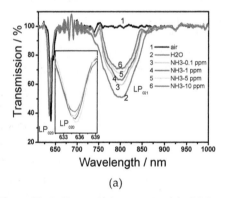 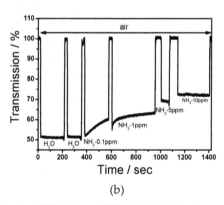

(a)   (b)

Figure 13. (a) Transmission spectra of the LPG coated with a TSPP infused $(PDDA/SiO_2)_{10}$ film due to immersion into water and into ammonia solutions of different concentrations: "H2O", LPG exposed into water; "air", LPG in air after drying with N2 gas; "NH3 x ppm", LPG exposed into a x ppm ammonia solution, where x = 0.1, 1, 5 and 10. (b) Dynamic response to water and ammonia solutions (0.1, 1, 5 and 10 ppm) recorded at 800 nm; LP_{020} and LP_{021} are labelling the linear polarized 020 and 021 modes, respectively.

The response of the transmission spectrum to varying concentration of ammonia is shown in Figure 13a. The dynamic response of the sensor was assessed by monitoring the transmission at the centre of the LP_{021} resonance band at 800 nm. The response is shown in Figure 13b, where "air" region and "H2O" and "NH3" regions correspond to the transmission recorded at 800 nm after drying the LPG and exposing the devise into water and ammonium solutions, respectively. After repeating the process of immersion in water and drying 4 times, the recorded spectrum was stable, demonstrating the robustness and stability of the employed molecules in aqueous environments (H_2O regions indicated in Figure 13). On immersion in 1 ppm and 5 ppm ammonia solutions, the transmission measured at 800 nm increases. The transmission when the coated LPG was immersed in a 10 ppm ammonia solution exhibits a further increase, reaching a steady state within 100 s, as shown in Figure 13b. The resonance feature corresponding to coupling to the LP_{020} cladding mode exhibits additional small red shifts of 0.5 and 1.5 nm when subsequently immersed in solutions of 1 ppm and 10 ppm ammonia concentration, respectively, along with decreases in amplitude, as shown in Figure 13a. The limit of detection (LOD) for the 100 μm period LPG coated with a $(PDDA/SiO_2)_{10}$ film that was infused with TSPP was 0.14 ppm and 2.5 ppm when transmission and wavelength shift were measured respectively. The LOD was derived from the calibration curve and the following equation [36, 44].

$$LOD = 3 \cdot \sigma / m \tag{2}$$

where σ is the standard uncertainty obtained as a symmetric rectangular probability [45]; m is the slope of the calibration curve. The sensing mechanism postulated is based upon the UV-vis spectrometer, QCM and LPG fibre measurements, and can be illustrated using Figure 14. As mentioned previously, while most of TSPP molecules form J-aggregates in the $PDDA/SiO_2$ film, some of the molecules are present in monomeric forms, as shown in Figure 14a, which may be easily sulfonated into species with neutral ($TSPP^{4-}$) and protonated (H_2TSPP^{2-}) pyrrole rings in water. As can be concluded from UV-vis measurements, on immersion of the TSPP-infused $PDDA/SiO_2$ film into water, most of J-aggregated TSPP molecules are removed from the mesopores between SiO_2 NPs, Figure 12. This indicates that the intermolecular interaction between J-aggregates of TSPP can be easily broken in water. However, some strongly bond TSPP compounds remained in the porous film. Figure 14b shows the chemical reactions involved in the interaction of ammonia with the $TSPP^{4-}$ and H_2TSPP^{2-} monomers. The electrostatic interaction between TSPP and PDDA is disturbed by the formation of ammonium ions and this causes the further desorption of the TSPP compound from the $PDDA/SiO_2$ film and consequently a decrease in the refractive index of the film.

It is important to consider the influence of the pH of the ammonia solutions on the sensor response, as the electrostatic interaction between TSPP and PDDA can be disturbed by OH^- ions in solution. To check this, the pH of the ammonia solutions used in this work were measured using a compact pH meter (B-211, Horiba), showing 7.3, 7.3 and 7.6 for 0.1, 1, and 10 ppm solutions, respectively. Thus, the molar concentration of NH_3 and OH^- is estimated to be 0.58, 5.8 and 58 mM for NH_3 and 0.25, 0.25 and 50 nM for OH^-, respectively. These data reveal that the concentration of OH^- does not have a significant role and the sensing mechanism is mainly based on the basicity of ammonia, as shown in Figure 14b. The cross sensitivity of the LPG sensor was tested using ethanol and methanol aqueous solutions. There was no measurable response of the sensor at the concentration levels similar to those tested for ammonia (0.1, 1, 10 and 100 ppm) indicating high selectivity of the sensor device to ammonia over those analytes. At much higher concentrations (10,000 ppm), however, blue-shift of the LP_{020} band and decrease in transmission at 800 nm was registered, which can be ascribed to the change of the bulk RI of the solution [6].

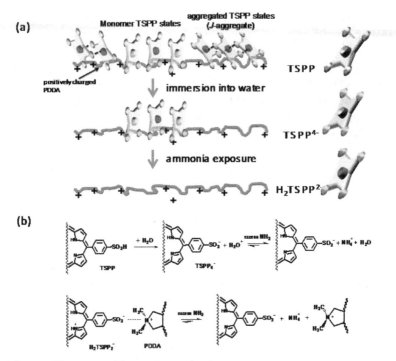

Figure 14. Schematic illustration of the sensing mechanism.

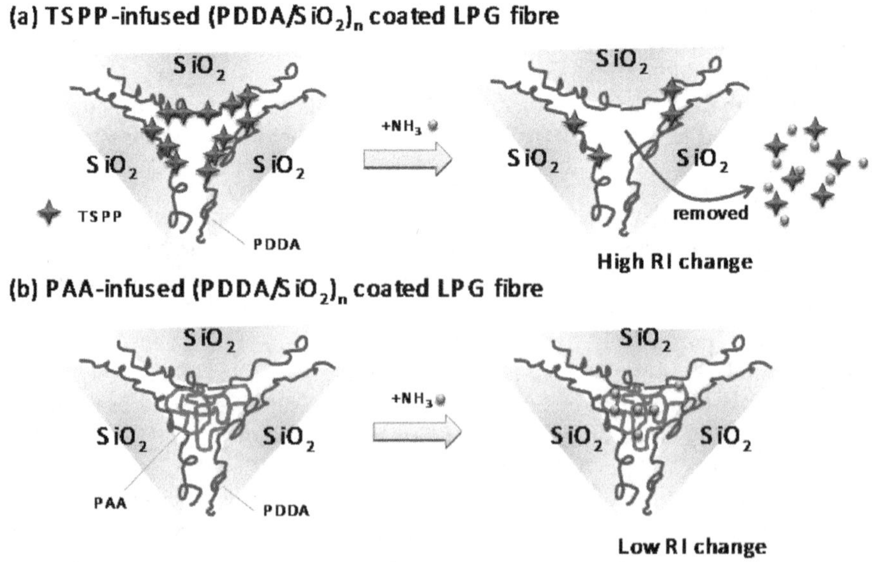

Scheme 1. Schematic illustration of the ammonia sensing mechanism for the LPG fibres modified with (a) TSPP-infused PDDA/SiO₂ and (b) PAA-infused PDDA/SiO₂ films.

The assessment of the cross sensitivity to other amine containing compounds is in progress. When compared with other ammonia optical sensors [46] the developed LPG device shows similar detection levels to coulorometric [47] and absorption spectroscopy [48] devices. In particular, the current LPG optical fibre sensor modified with a mesoporous thin film offers unique advantages such as versatile chemical infusion of various chemicals into the mesopores, fast response time owing to the easy analyte penetration and robustness. In addition, due to unique properties of the optical fibre such as biocompatibility, multiplexing, small size, immunity to electromagnetic interference and possibility to work in harsh environment, a cost effective, portable sensor system can be produced. To further elucidate the sensing principle of the LPG sensor, the response of a $(PDDA/SiO_2)_{10}$ coated LPG and of a PAA-infused $(PDDA/SiO_2)_{10}$ coated LPG was further examined. There was no change when the $(PDDA/SiO_2)_{10}$ and PAA infused $(PDDA/SiO_2)_{10}$ coated LPGs were exposed to ammonia solutions which have similar values of RI, 1.3329, from 1 ppm to 100 ppm concentrations as measured using a portable refractometer (R-5000, Atago). Thus, no change in sensor signal was recorded in the $(PDDA/SiO_2)_{10}$ coated LPG. On the other hand, higher affinity to ammonia based on the acid-base interaction is expected in the PAA-infused $(PDDA/SiO_2)_{10}$ coated LPG [39]. However, no response to ammonia was observed. This indicates that the adsorption of ammonia in small molecular size results in a small RI change. Consequently, the selective desorption of the functional compound with a high RI from the mesoporous film results in a significant increase in the device sensitivity, as shown in Scheme 1. This measurement principle can be further explored for a different set of analyte-functional compound pair including biological analytes expanding the application range of the proposed LPG device.

6. Device repeatability

It was observed that, if the sensor was repeatedly exposed to a certain concentration of ammonia following a washing and drying cycle, the response was not reproducible, in that on each subsequent exposure to ammonia the extinction of the band was further reduced, as shown in Figure 13. The extinction of the band still changes in time and the effect saturates with the increase of the ammonia concentration, in the way indicated in Figure 13 b. The linear dependence of the sensor response upon the ammonia concentration indicates that magnitude of the change on each exposure is the same; thus proposed device can be employed for measurements of the cumulative exposure to ammonia. However, after a number of repeated exposures, with the number being dependent on the concentration of the ammonia solution to which the device was exposed, the sensitivity was exhausted, and exposure to the ammonia solution would produce a spectrum equivalent to that obtained after deposition of the porous film, but before TSPP infusion [5]. It was found that the TSPP compound could be infused into the porous film again by the exposure of the LPG to a 1 mM solution of TSPP (step (vii) in Figure 5). The reproducibility of the device was tested by the exposure of the LPG coated with the TSPP infused $(PDDA/SiO_2)_{10}$ film to an ammonia solution of concentration >1000 ppm (concentration chosen to ensure that the TSPP was completely desorbed from the PDDA/SiO2 film), followed by washing with pure water and immersion in a 1 mM TSPP solution in order to regenerate the original film properties [5], Figure 11c. The procedure was repeated 5 times and the same behaviour and resulting LPG transmission spectrum were obtained after each reinfusion process. It should be noted that ammonia measurements were conducted 3 times at each concentration. These results indicate that highly reproducible measurements can be conducted by employing TSPP reinfusion step.

7. Summary

The response of the transmission spectrum of an LPG of period 100 μm to the deposition of a multilayer film of SiO_2 NPs and the subsequent infusion of a porphyrin into the porous coating has been characterized. The infusion of the functional materials, chosen to be sensitive to the analyte of interest, into the base mesoporous coating was reported. Two possible sensing mechanisms have been exploited, based upon changes in the refractive index of the coating resulting from (1) chemically induced RI changes of the mesoporous coating at the adsorption of the analyte to the functional material, namely PAA and (2) chemically induced desorption of the functional material, namely TSPP, from the mesoporous coating. The operation of the device as a re-useable ammonia sensor with a minimum detection level of 0.14 ppm and a response time of approximately 100 s exploiting desorption sensing mechanism has been reported. On the other hand, the ammonia adsorption to the carboxylic functional groups of the PAA resulted in a small RI change and low sensitivity to analyte. The film thicknesses and functional material infusion time employed in this work have been determined empirically, as the calculation of the RI of the porous coating infused with TSPP and immersed in water would be highly complex. Operation of the system at the point of coincidence of the mode transition region and phase matching turning point for films of larger/smaller thickness could be achieved by decreasing/increasing the quantity of the function dye infused into the film, and this may influence the minimum detectable concentration and sensitivity. Such issues are currently under investigation. Advantages of the proposed method lie in the ability to control functionality of the coating by, for instance, choosing different matrix polymers that extend the class of the detectable analytes. Additionally, infusion of different types of functional compounds would allow the detection of different chemicals using a similar principle of operation. Future work is planned to demonstrate the current system for sensing a variety of chemical and biological compounds and for gas sensing.

Author details

Sergiy Korposh and Seung-Woo Lee[*]
Graduate School of Environmental Engineering, the University of Kitakyushu, 1-1 Hibikino, Wakamatsu-ku, Kitakyushu, Japan

Stephen James and Ralph Tatam
School of Engineering, Cranfield University, Cranfield, Bedford, UK

---

[*] Corresponding Author

Acknowledgement

This work was supported by the Regional Innovation Cluster Program of the Ministry of Education, Culture, Sports, Science and Technology (MEXT), Japan and partly by the Ministry of Knowledge Economy (MKE, Republic of Korea) via the Fundamental R&D Program for Core Technology of Materials. The authors from Cranfield are grateful to the Engineering and Physical Sciences Research Council, EPSRC, UK for funding under grants EP/D506654/1 and GR/T09149/01.

8. References

[1] Tsuda H, Urabe K. Characterization of long-period grating refractive index sensors and their applications. Sensors 2009; 9 4559-4571.

[2] Libish T M, Linesh J, Biswas P, Bandyopadhyay S, Dasgupta K, Radhakrishnan P, Fiber Optic Long Period Grating Based Sensor for Coconut Oil Adulteration Detection. Sensors & Transducers Journal 2010; 114 102-111.

[3] James S W, Tatam R P. Optical fibre long-period grating sensors: characteristics and application. Measurement Science and Technology 2003; 14 R49-R61.

[4] Martinez-Rios A, Monzon-Hernandez D, Torres-Gomez I. Highly sensitive cladding-etched arc-induced long-period fiber gratings for refractive index sensing. Optics Communications 2010; 283 958-962.

[5] Korposh S, James S W, Lee S-W, Topliss S M, Cheung S C, Batty W J, Tatam R P. Fiber optic long period grating sensors with a nanoassembled mesoporous film of SiO2 nanoparticles. Optics Express 2010; 18 13227-13238.

[6] Korposh S, James S W, Tatam R P, S-W Lee. Refractive index sensitivity of fibre optic long period gratings coated with SiO2 nanoparticle mesoporous thin films. Meas. Sci. Tech. 2011; 22 075208s.

[7] Schroeder K, Ecke W, Mueller R, Willsch R, Andreev A A. A fibre Bragg grating refractometer. Measurement Science and Technology 2001; 12 757-764.

[8] Asseh A, Sandgren S, Åhlfeldt H, Sahlgren B, Stubbe R, Edwall G. Fiber optical bragg grating refractometer. Fiber and Integrated Optics 1998; 17(1) 51-62.

[9] Iadicicco A, Campopiano S, Cutolo A, Giordano M, Cusano A. Self temperature referenced refractive index sensor by non-uniform thinned fiber Bragg gratings. Sensors and Actuators, B: Chemical 2006; 120 231–237.

[10] Laffon G, Ferdinand P. Tilted short-period fibre-Bragg-grating-induced coupling to cladding modes for accurate refractometry. Measurement Science and Technology 2001; 12 765-770.

[11] Buggy S J, Chehura E, James S W, Tatam R P. Optical fibre grating refractometers for resin cure monitoring. Journal of Optics A: Pure and Applied Optics 2007; 9 S60-S65.

[12] Paladino D, Cusano A, Pilla P, Campopiano S, Caucheteur C and Mégret P. Spectral behavior in nano-coated tilted fiber Bragg gratings: Effect of thickness and external refractive index. IEEE Photonics Technology Letters 2007; 19 2051–2053.

[13] Caucheteur C, Paladino D, Pilla P, Cutolo A, Campopiano S, Giordano M, Cusano A, Mégret P. External refractive index sensitivity of weakly tilted fiber Bragg gratings with different coating thicknesses. IEEE Sensors Journal 2008; 8 1330-1336.

[14] V. Bhatia V, Vengsarkar A M. Optical fibre long-period grating sensors. Optics Letters 1996; 21 692-694.

[15] Shi Q, Kuhlmey B T. Optimization of photonic bandgap fiber long period grating refractive-index sensors. Optics Communications 2009; 282 4723-4728.

[16] Davies E, Viitala R, Salomäki M, Areva S, Zhang L, Bennion I. Sol-gel derived coating applied to long-period gratings for enhanced refractive index sensing properties. Journal of Optics A: Pure and Applied Optics 2009; 11 art. no. 015501.

[17] Allsop T, Webb D J, Bennion I. A comparison of the sensing characteristics of long period gratings written in three different types of fiber. Optical Fiber Technology 2003; 9 210-223.

[18] Yu X, Shum P, Ren G B, Ngo N Q. Photonic crystal fibers with high index infiltrations for refractive index sensing. Optics Communications 2008; 281 4555-4559.

[19] Zhu Y, He Z, Kaňka J, Du H. Numerical analysis of refractive index sensitivity of long-period gratings in photonic crystal fiber. Sensors and Actuators, B: Chemical 2008; 129 99-105.

[20] Shi Q, Kuhlmey B T. Optimization of photonic bandgap fiber long period grating refractive-index sensors. Optics Communications 2009; 282 4723-4728.

[21] Rees N D, James S W, Tatam R P, Ashwell G J. Optical fiber long-period gratings with Langmuir- Blodgett thin-film overlays. Optics Letters 2002; 27 686-688.

[22] Allsop T, Floreani F, Jedrzejewski K P, Marques P V S, Romero R, Webb D J, Bennion I. Spectral characteristics of tapered LPG device as a sensing element for refractive index and temperature. Journal of Lightwave Technology 2006; 24 870-878.

[23] Iadicicco A, Campopiano S, Giordano M, Cusano A. Spectral behavior in thinned long period gratings: effects of fiber diameter on refractive index sensitivity. Applied Optics 2007; 46 6945-6952.

[24] Martinez-Rios A, Monzon-Hernandez D, Torres-Gomez I. Highly sensitive cladding-etched arc-induced long-period fiber gratings for refractive index sensing. Optics Communications 2010; 283 958-962.

[25] Ishaq I M, Quintela A, James S W, Ashwell G J, Lopez-Higuera J M, Tatam R P. Modification of the refractive index response of long period gratings using thin film overlays. Sensors and Actuators, B: Chemical 2005; 107 738-741.

[26] Smietana M, Korwin-Pawlowski M L, Bock W J, Pickrell G R, Szmidt J. Refractive index sensing of fiber optic long-period grating structures coated with a plasma deposited diamond-like carbon thin film. Measurement Science and Technology 2008; 19 art. no. 085301.

[27] Rees N D, James S W, Tatam R P, Ashwell G J. Optical fiber long-period gratings with Langmuir- Blodgett thin-film overlays. Optics Letters 2002; 27 686-688.

[28] Cusano A, Iadicicco A, Pilla P, Contessa L, Campopiano S, Cutolo A, Giordano M. Cladding mode reorganization in high-refractive-index-coated long-period gratings: effects on the refractive-index sensitivity. Opt. Lett. 2005; 30 2536-8.

[29] Yang J, Yang L, Xu C-Q, Li Y. Optimization of cladding-structure-modified long-period-grating refractive-index sensors. Journal of Lightwave Technology 2007; 25 372-380.

[30] Del Villar I, Achaerandio M, Matias I R, Arregui F J. Deposition of overlays by electrostatic self assembly in long-period fibre gratings. Opt. Lett. 2005; 30 720-722.

[31] Del Villar I, Matias I R, Arregui F J. Influence on cladding mode distribution of overlay deposition on long period fiber gratings. J. Opt. Soc. Am. A 2006; 23 651-658.

[32] Cheung S C, Topliss S M, James S W, Tatam R P. Response of fibre optic long period gratings operating near the phase matching turning point to the deposition of nanostructured coatings. J. Opt. Soc. Am. B 2008; 25 897-902.

[33] Corres J M, Matias I R, del Villar I, Arregui F J. Design of pH sensors in long-period fiber gratings using polymeric nanocoatings. IEEE Sens. J. 2007; 7 455-463.

[34] Keith J, Hess L C, Spendel W U, Cox J A, Pacey G E. The investigation of the behavior of a long period grating sensor with a copper sensitive coating fabricated by layer-by-layer electrostatic adsorption. Tatanta 2006; 70 818-822.

[35] Schick G A, Schreiman I C, Wagner R W, Lindsey J S, Bocian D F. Spectroscopic characterization of porphyrin monolayer assemblies. J. Am. Chem. Soc. 1989; 111 1344-1350.

[36] Korposh S, Selyanchyn R, Yasukochi W, Lee S-W, James S, Tatam R. Optical fibre long period grating with a nanoporous coating formed from silica nanoparticles for ammonia sensing in water. Materials Chemistry and Physics 2012; 133 784-792.

[37] Lee S-W, Takahara N, Korposh S, Yang D-H, Toko K, Kunitake T. Nanoassembled thin film gas sensors. III. Sensitive detection of amine odors using TiO2/poly(acrylic acid) ultrathin film quartz crystal microbalance sensors. Anal. Chem., 2010; 82 2228-2236.

[38] Ye C C, James S W, Tatam R P. Simultaneous temperature and bend sensing with long-period fiber gratings. Optics Letters 2000; 25 1007-1009.

[39] Bravo J, Zhai L, Wu Z, Cohen R E, Rubner M F. Transparent superhydrophobic films based on silica nanoparticles. Langmuir 2007; 23 7293-7298.

[40] Currie E P K, Sieval A B, Fleer G J, Cohen Stuart M A. Polyacrylic acid brushes: Surface pressure and salt-induced swelling. Langmuir 2000; 16 8324-8333.

[41] Korposh S, Kodaira S, Batty W J, James S W, Lee S-W. Nano-assembled thin film gas sensor. II. An intrinsic high sensitive fibre optic sensor for ammonia detection. Sensor and Materials. 2009; 21(4) 1790-189.

[42] Korposh S O, Takahara N, Ramsden J J, Lee S-W, Kunitake T. Nano-assembled thin film gas sensors. I. Ammonia detection by a porphyrin-based multilayer film. J. Biol. Phys. Chem. 2006; 6 125-133.

[43] Kadis KM, Smith KM, Guilard R. The Porphyrin Handbook. vol. 11–20, 3500. Academic Press, 2003.

[44] Swartz ME, Krull IS. Analytical Method Development and Validation, Marcel Dekker, Inc.: NY USA; 1997.

[45] Rafael G, Possetti C, Kamikawachi R C, Muller M, Fabris J L. Metrological evaluation of optical fiber grating-based sensors: An approach towards the standardization. J. Lightwave Technology (OFS-21). (2011) 2167500, DOI 10.1109/JLT.2011.2167500.

[46] Timmer B, Olthuis W, Berg A. Ammonia sensors and their applications—a review. Sens. Act. B. 2005; 107 666-677.

[47] Yimit A, Itoh K, Murabayashi M. Detection of ammonia in the ppt range based on a composite optical waveguide pH sensor. Sens. Act. B. 2003; 88 239-245.

[48] Huszár H, Pogány A, Bozóki Z, Mohácsi Á, Horváth L, Szabó G. Ammonia monitoring at ppb level using photoacoustic spectroscopy for environmental application. Sens. Act. B. 2008; 134 1027-1033.

Chapter 6

Photonic Sensors Based on Flexible Materials with FBGs for Use on Biomedical Applications

Alexandre Ferreira da Silva, Rui Pedro Rocha,
João Paulo Carmo and José Higino Correia

Additional information is available at the end of the chapter

1. Introduction

A wide variety of optical fiber sensors are available and can be divided into three categories: the external or extrinsic ones (Beard P. C. et al., 1996) where the fiber is only used to drive the measured information to and from the transducer at a distant location, the intrinsic category (Boerkamp M. et al., 2007) where the optical properties are sensitive to an external stimulus (Grattan S. K. T. et al., 2009; Gu X. et al., 2006), and the hybrid category where the light is transferred over the optical fiber for conversion into electricity on a distant optical receiver (Yao S.-K. et al., 2003).

From the previously mentioned categories, the intrinsic sensors, where FBGs are included, have been studied and applied intensively during the past 20 years (Lee B., 2003). The Bragg grating structure is the intrinsic element to the fiber responsible for the sensor behavior. The gratings can be inscribed by ultraviolet (UV) light beams, taking advantage of the optical fiber photosensitivity (doped with germanium) to this radiation. In addition to the standard advantages attributed to the optical fiber sensors, FBGs have an inherent self-referencing and multiplexing capability. Essentially, the FBG is a periodic variation of the refraction index along the fiber axis. As illustrated in the Figure 1, this structure works as a reject-band filter, reflecting back the spectral component, λ_B [nm], which satisfies the Bragg condition (given by equation (1)) and transmitting the remaining components. The Bragg wavelength is given by (Hill K. O. et al., 1997):

$$\lambda_B = 2n_{eff}\Lambda \tag{1}$$

where Λ [nm] is the grating pitch and n_{eff} is the effective refraction index of the fiber core. The wavelength shift, $\Delta\lambda_B$ [nm], of a FBG sensor subject to a physical disturbance is given by (Wei C.-L. et al., 2010):

$$\frac{\Delta\lambda_B}{\lambda_B} = (1 - \rho_e)\Delta\varepsilon + (\alpha + \xi)\Delta T \tag{2}$$

where ρ_e, $\Delta\varepsilon$, α, ξ, and ΔT are the effective photoelastic constant, the axial strain, the thermal expansion, the thermal optic coefficient and the temperature shifts, respectively. The ratio in the first term of equation (2) expresses the strain effect on an optical fiber. It corresponds to a change in the grating spacing and the strain-optic induced change in the refractive index. The temperature sensing is mainly related with the second term of the expression. As the FBG is subjected to temperature variation, it dilates or contracts, modifying the grating pitch.

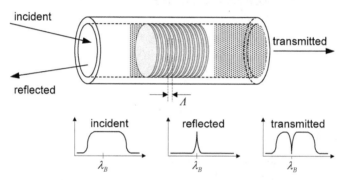

Figure 1. Illustration of working principle of FBGs.

One main advantage of this system is the measurements made on the wavelength instead of optical power. This enables a system that is not sensitive to external factors as fluctuations of the optical source. The stability is also extended to the bond between the polymer matrix and the optical fiber in which it is wrapped. Therefore, these features make the FBGs suitable sensing elements for doing physical measurements, where a kind of displacement is available. Examples of such applications found in the literature include the measurements of strain (Grattan S. K. T. *et al.*, 2009; Ling H. Y. *et al.*, 2006), pressure (Peng B. J. *et al.*, 2005; Zhang W. et al., 2009), force (Rajan G. *et al.*, 2010; Zhao Y. *et al.*, 2005), tilt rotation by an angle (Peng B. J. *et al.*, 2006; Xie F. *et al.*, 2009), acceleration (Antunes P. *et al.*, 2011; Fender A. *et al.*, 2008), temperature (Bao H. *et al.*, 2010; Gu X. *et al.*, 2006), humidity (Arregui F. J. *et al.*, 2002; Yeo T. L. *et al.*, 2005), magnetic fields (Orr P. *et al.*, 2010), cardiorespiratory function (Silva A. F. *et al.*, 2011a), hand posture analysis (Silva A. F. *et al.*, 2011b), gait function analysis (Rocha R. P. *et al.*, 2011) and integration on wearable garments (Carmo J. P. *et al.*, 2012).

This chapter focuses on biomedical applications of FBGs embedded into flexible carriers for enhancing the sensitivity and protection to the optical fiber, and to provide interference-free instrumentation. The same FBG system was used in all experiments presented in this chapter. In terms of construction, this FBG system is composed by a sensing and a monitoring module. Figure 2 shows photographs of the light source and the hardware used in the interrogation system for monitoring the received light which is then seen on a computer screen. The Fiber-Bragg Grating used in these experiments was produced by the FiberSensing company (FiberSensing, 2012). The grating is 8 mm long with a resonance wavelength in the 1550 nm range, which corresponds to a refraction index modulation period of the core in the half-micrometer range. The interrogation monitor (I-MON 80D from Ibsen Photonics company (Ibsen, 2012) allows real-time spectrum monitoring of FBG sensors interrogation systems. Along with the interrogation monitor, software is supplied by the manufacturer that permits real-time visualization of the waveforms while the sensor is being actuated. This system has a resolution of 10 pm.

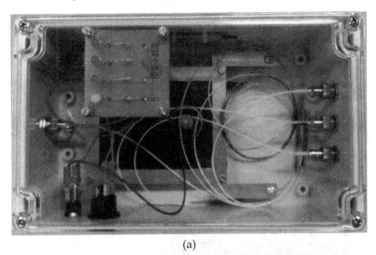

(a)

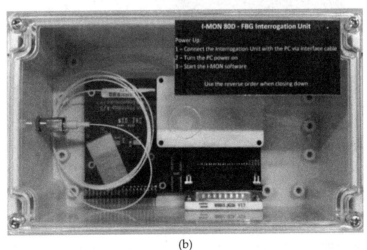

(b)

Figure 2. Proposed system's non-sensing parts composed by (a) broadband light source (Denselight, 2012) and in (b) the optical circulator (Oplink, 2012) and interrogation monitor hardware (Ibsen, 2012).

A carrier material made of polychloroethanediyl (polyvinyl chloride, or simply PVC) was used in the FBG embedment for increasing their sensitivity to strains and at the same time to improve the adhesion to the surface under measurement (Silva A. F. et al., 2012). The PVC material was selected as FBG carrier due to its excellent performance/cost ratio and easy handling during the manufacturing process. Moreover, the PVC presents many other advantages when compared with its direct competitors (e.g., either the polyurethane or the polyolefin) such as low production cost, making this material highly competitive. On the practical side, it offers high resistance to aging, high versatility and simplicity of maintenance (Silva A. F. et al., 2012). The cross-sections of the Figure 3 show the configuration of the layers within the carrier (few photographs was taken under different directions and illuminations for better illustrating the FBGs and layers that constitute the carrier).

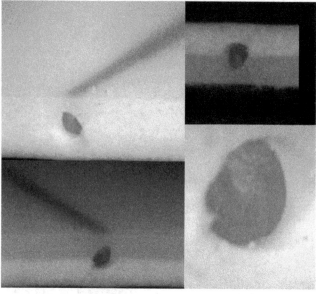

Figure 3. Few photographs showing views with the cross-section of the three layers that constitutes the carrier. The supported FBGs are also showed.

2. Knee's kinematic monitoring

2.1. Introduction

In this section of the chapter is presented a sensing electronic-free wearable solution for monitoring the knee-referenced gait process as a biomedical application example using Fiber Bragg Grating (FBGs) sensors. This sensing system is based on a single optical FBG, with a resonance wavelength of 1547.76 nm, which shifts to lower or higher wavelengths when subjected to strain variations, with a resolution of 10 pm. The measuring of the knee movements, flexion and extension with the corresponding joint acting as the rotation axis, is shown for a healthy individual. The optical fiber with the FBG is placed inside a polymeric foil (composed by three flexible layers), attached to an elastic knee band, which facilitates its placement in the knee (centered in the patella) while maintaining full sensing capabilities. Although the knee is used here as the example, the way the device is placed on the specific body part to be measured enables the clear detection of the movements in respect to the corresponding joint. The proposed prototype was evaluated under different condition tests and also to assess its consistency and flexibility of use. The designed sensor demonstrates advantages in biomedical fields such as physical therapy and athletic assessment applications because of the system's resolution and easiness of applying it onto the body part under investigation. Another advantage is the possibility to measure, record and evaluate specific mechanical parameters of the limbs' motion. Patients with bone, muscular and joint related health conditions, as well as athletes, are within the most important end-user applications. Moreover, this system can be used simultaneously with, for example, inertial and magnetic sensors enabling the correlation between the measured wavelengths with angular degrees.

During the past years, body kinematics monitoring in human beings is a growing area within the field of engineering applied to medicine. Universities, high-performance sport centers and health-care institutions have been developing ways to accurately measure and evaluate the way the human body moves for endless purposes. The main objectives for such attention in measuring and evaluating the human body kinematics are improvements of athletic performance (Anderson D. et al., 1994; Yamamoto Y., 2004; von Porat A. et al., 2007) in competitions and historic evaluation studies of patients to determine if the prescribed therapy is being efficient and evaluating the rehabilitation of patients (Yang X. J. et al., 2012; Vancampfort D. Et al., 2012; Cup E. H. et al., 2007; Kun L. et al., 2011) based on the information provided by measuring the limbs' movements. Several systems for body kinematics monitoring have been realized using different approaches such as complex electronic systems including a 2.4 GHz radio-frequency (RF) transceiver (Afonso J. A. et al., 2010), motion capture techniques (Ren L. et al., 2008; Parker T. M. et al., 2008) and advanced software algorithms that demand profound specific know-how and are also very complex (Moustakidis S.P. et al., 2010; Wu Y. et al., 2011). Other applications for limb posture monitoring include the assessment of certain neurologic and orthopedic diseases (Yavuzer G. et al., 2008; Mavrogiorgoua P. et al., 2001; Turcot K. et al., 2008). A gait monitoring system based on optical fiber, complemented with a motion capture system, has already been proposed but, when compared to the solution presented in this paper, it shows several differences including: use of a plastic optical fiber (POF), calibration procedure required and measurement based on the transmitted optical power when the POF is bent (Bilro L. et al., 2011). Therefore, the importance of measuring (Godfrey A. et al., 2008) and characterizing the limbs' kinematics is quintessential in diagnosing physical and mental disorders, originated by trauma, stroke or disease, and determining the appropriate treatment and therapy. The data can be saved (using the setup showed in the Figure 2) for further analysis and study which enables comparisons between results to be made along the time. Moreover, the proposed system was designed to obtain maneuverability making it compatible with free range body kinematics movements.

2.2. Approach

The gait cycle can be defined as the sum of the two components that compose a full step, *e.g.* the stance and the swing phases. The stance and swings phases comprehend the periods when the foot is touching the ground and advancing in the air permitting the progression of

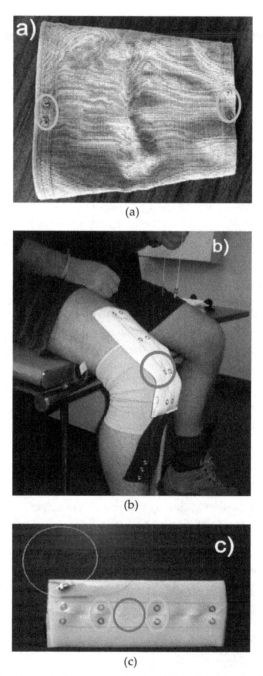

Figure 4. In (a), the elastic knee with the pressure buttons signaled with green ellipses. In (b), the sensing part attached to a standard elastic knee band and in (c), a close-up just of the PVC foil with the embedded FBG signaled with a red circle and the pressure buttons with green ellipses.

the body, respectively. The knee kinematics is represented by two stages: flexion and extension. The objective is to represent graphically, as a function of the measured wavelength, the full human gait period, centered on the knee joint using just one FBG and a single mode optical fiber. This section focus on the validation of the proposed concept by measuring the knee's kinematics, the single fiber and single FBG sensor, placed in the center of the knee (patella), are enough to measure and evaluate the subject's evolution. In order to make this possible, a high-sensitivity sensor is necessary to detect the full movement from one extreme (when the leg is completely straight) to the other (maximum knee deflection during gait) and all movements that happen in-between, i.e., stance and swing phases. The sensing part is based on a flexible structure that can be placed/removed on/from the knee very easily. This is done by using small pressure buttons as attaching elements. The Figure 4 shows in more detail the small metallic pressure buttons that attach the different components of the sensing system, the elastic knee band already placed and the foil with the embedded FBG. This type of elastic knee band is regularly used in prevention/precaution situations in people with a temporary or permanent muscular injury enabling the use of the flexible structure by any person and in any junction in the body. The pressure buttons ensure that the sensing element is able to sense the flexion and extension of the knee as the subject moves around. Since optical fibers are immune to electromagnetic interference (EMI) and can be used safely in wet environments or even under water, the proposed solution increases the number of possible applications for this technology.

2.3. Flexible sensing structure

The accurate measurement of the knee joint movement is possible if the dynamic range of the sensor is increased. This can be accomplished by proper selection of the substrate material that can conform correctly to the actual movement. Therefore, a structure with enough area to cover the knee enables the transference of the movements to the embedded sensor. A wide rectangular configuration was chosen to cover both the flexion and the extension movements since it provides the required area of contact to be translated by the sensing area and allows the light to travel without any abrupt corners that would obstruct the communication with the monitoring stage. The main characteristics, and advantages, of this foil include flexibility, stretchability and the capability to sustain a good bonding between the optical fiber and the substrate. The host material is polyvinyl chloride (PVC) with custom formulation to assure the bonding and the stimulus transfer (Silva A. et al., 2012). Moreover its size and shape are completely customizable during fabrication.

2.4. Samples and experiments

As previously stated, the knee kinematics is characterized mainly by the flexion and extension dynamics. In order to monitor these movements, the flexible polymeric foil prototype, having the sensor embedded in it, was applied to an elastic knee band placing the FBG correctly on the patella using pressure buttons as bonding elements. The Figure 5 shows the raw data measured by the FBG with the volunteer walking and running on top of a commercially available treadmill validating the system's consistency and reliability.

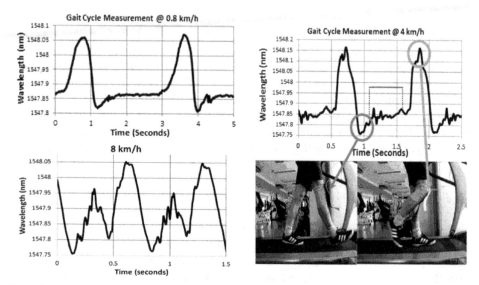

Figure 5. FBG's measured raw unfiltered two full gait cycles for three different speeds. Key points of the swing and stance phases are represented by the minimum and maximum deflections of the FBG, and the time in-between them, respectively.

The periodicity of the signals and the respective association with the gait cycle are easily recognized, *i.e.*, all the different movements associated with a full step are clearly identifiable in the measured waveforms. The tests were done during 10 seconds but only two full gait cycles are considered for clear visualization of the acquired raw unfiltered data. In the Figure 5 it is seen the FBG's raw data waveforms for three different walking speeds, as an example of the system's capability, at 0.8 km.h^{-1} (0.22 m.s^{-1}), 4 km.h^{-1} (1.11 m.s^{-1}) and 8 km.h^{-1} (2.22 m.s^{-1}). For the highest speed, the measured curve becomes sharper with few differences. The running movement is a quicker step which leads to the waveform sharpness and presents a slight different kinematics, mainly when the front foot reaches the ground and pulls the entire body forward. Nonetheless, the results enable the identification of the several events occurring during the walking and running movements. In order to compare the period of the measured data with the different stages of a full step, the Figure 5 shows the two extreme values obtained by the system during a complete step on a 2.5 seconds measurement at 4 km.h^{-1}. The stage where the flexion of the leg in the knee joint, hence on the FBG as well, is the minimum possible during a full step is visible corresponding to a value of 1547.76 nm (resonance wavelength). The maximum deflection obtained during a step corresponds to a value of 1548.16 nm. Between the two extremes of the movement, there is a stage with low amplitude variations that follows the knee movement while the leg is in contact with the floor. It should be noted that between the minimum and maximum values, the leg is always touching the floor (stance phase) and the FBG sensor is basically completely stretched with just minor variations related to the elastic knee band used. This relatively constant period is marked in red in the chart seen in the Figure 5. The right leg starts its movement backwards touching the floor first with the ankle, proceeding with the base of the foot until it begins the movement forward with the tip of the toes (the maximum deflection represented in the figure represents around 80% of the gait period (Trew M. et al., 2005). The FBG based sensor enables identification of all the different movements associated with a full step allowing comparison between different results acquired in different situations. The presented solution for the knee can also be applied to any other joint in the human body.

2.5. Discussion

The measurements seen in Figure 3 show a smooth waveform at low speeds. For speeds above 4 km/h the elasticity factor of the knee band and the slip occurring during the movement change slightly the position of the attached FBG in the knee-band. It can be concluded that the elastic band matches perfectly the skin for slow movements (roughly≤4 km.h^{-1}) guaranteeing that the FBG is always on the correct place. Due to the FBG's high resolution, 10 pm, the slightest slip induces immediately a change in the output data. This means that a "calibration" procedure is needed to guarantee that the sensor is exactly on top of the patella. Also, the elastic band used is specified for a leg perimeter around the knee of 35-38 cm. For the subjects with a corresponding leg girth, it reproduces correctly the knee movement during the gait. For the remaining ones, a slight displacement of the FBG occurs due to the vibration induced in the leg while taking steps forward, and therefore explaining the wavelength variations observed in the measurements. Typically, the knee position tends to be steady at low speeds as seen in the test performed at 0.8 km.h^{-1} in the Figure 5, but its variations become more significant as the speed increases. The stronger vibration caused by the running steps is the source for the somewhat abrupt peaks in the waveform at 8 km.h^{-1}. Different sizes of the elastic knee band and the way it is attached to the foil with the embedded sensor or even embed it in the textile (Grillet A. et al., 2008) of an elastic knee band can also reduce these fast oscillations. Regarding the sensing electronic modules, the fast changes observed can also be explained.

2.6. Conclusions

A structure made of polyvinyl chloride (PVC) material, carrying an embedded Fiber Bragg Grating (FBG) sensor with a 10 pm resolution, was attached to an elastic knee band. A clear characterization of the movement of the knee joint as a function of the wavelength variation and the associated angle measured between the tibia and femur were obtained. All the different movements associated with a full step, the stance and swing phases and their characteristic progression, are clearly identifiable in the obtained waveforms allowing comparison between different results acquired in different situations. The presented prototype is easy to connect and does not require technical personnel to give support and expertise making this approach very interesting as a functioning system for body kinematics monitoring. Moreover, since optical fiber is immune to electromagnetic interference and can support wet environments, including under water, the developed system opens new applications for body kinematics monitoring when a direct and easy relation between wavelength variation and angles is achieved. The FBG really demands a careful placement, and for walking speeds above 4 km.h^{-1}, the knee band slips and does not reproduce as accurately the gait cycle. Another solution for this problem could be to embed the optical fiber with the sensor in the actual textile to be used on the knee. The integration of a single optical fiber in a polymeric foil made of PVC resulted in a structure with a very good sensitivity for transducing accurately the knee flexion and extension during the walking and running tests. It is easy to install, comfortable to wear and accurately measures the body kinematics. Since this approach uses a flexible structure it can be worn by any person and it can be applied to other articulations as the shoulder or the elbow.

3. Hand posture monitoring

Following the knee's kinematics monitoring example, the FBGs can be used in more tricky examples as the hand posture monitoring. In reality, the FBG is a powerful tool for monitoring the body articulations due to its sensitivity, response and inherently properties as multiplexing.

The hand is possibly on the most complex articulating system mainly due to the density of articulations per volume. Moreover, apart from the anatomic structure, the hand is a key element to perform the interface between the human and the world. One could imagine how difficult it would be to perform its daily duties without using any hand.

However, the hand impairment is more common that one could image, mainly driven by stroke. Just in 2010, 73.7 billion dollars (Lloyd-Jones D. et al., 2010) were mainly spent on rehabilitation programs required to minimize muscle spasticity or pain and to recover from impairment. It is on this very stage that FBGs can be a great tool not only to enhance the rehabilitation programs but also to minimize their costs.

3.1. Hands' kinematics

The hand movements can be simplified to flexion-extension (straightening of the fingers) and abduction-adduction (pulling fingers apart or towards each other).

In today's physical therapy sessions, the exercises focus mainly on finger passive range of motion, fist making, object pick-up, finger extension and grip strengthening. In reality, one is mainly performing flexion-extension movements, which is the most frequently performed movement on a daily basis. Furthermore, the abduction-adduction movement has a much lower amplitude compared to the flexion-extension ones.

Based on these exercises, the therapist looks for data related to grip and pinch strength, joint range of motion, and functional abilities (Dipietro L. et al., 2003). Their assessment provides key-information for diagnosis, rehabilitation program and treatment progress analysis.

The most common tactic to retrieve the required data is based on the measurement of the finger range of movements while the subject is grabbing different balls of different densities. For the range of movement is measured via a goniometer place on each finger joint which allied to the ball density enables the strength calculus. As one can imagine, such technique is prone to errors, due to parallax effect, and inappropriate use of the equipment. Furthermore, such technique does not enable a simultaneous measurements of the entire had range of motion. As a result, the therapist takes a significant amount of time to perform all the required measurements that end up to have associated errors (Dipietro L. et al., 2003).

Independently of the performed exercises, the subjectivity associated to the patient's evaluation by the therapist leads to non-conclusive assessment of the patient's motor capacity and consequently misdiagnosis. Furthermore, the existing solutions are not suited to dynamic measurements. This scenario opens the opportunity for the development of a wearable device capable of performing an online monitoring of the hand kinematics in a more efficient manner.

3.2. State-of-the-art

In a generic scenario, a set of sensors applied to a glove are able to retrieve data related to the hand posture, from which directly or indirectly, depending on the sensor system architecture, other measurands can be also retrieved, e.g. pinch strength, motion range, among others (Dipietro L. et al., 2003).

It is already possible to find some wearable solutions capable of monitoring the hand posture and retrieving the required data, few based on electrically conductive elastomer (Lorussi F. et al., 2003; Lorussi F. et al., 2005; Scilingo E. P. et al., 2003; Tognetti A. et al., 2006), accelerometers (Perng J. K. et al., n.d.), induction coils, (Fahn C.-S. et al., 2005) and hetero-core fiber optic sensor (Nishiyama M. et al., 2009), but none on FBGs.

However, the available solutions are quite complex (Silva A. F. et al., 2011b), namely because of non-linear responses from the sensor, fragility issues or complex methods for signal processing. Still, from the existing technologies and solutions, the ones based on optical fiber sensors offer the biggest potential, when looking for performance and wearability (Lee B., 2003).

A solution based on FBG sensors can accomplish a simpler device compared to the existent ones by working on the sensor system design.

3.3. Monitoring approach

The finger movements on the joint site induce strain, namely tensile on the upper-side and compressive on the bottom-side, considering the open hand the steady-state. By applying a FBG on the joint site and use it as a strain gauge, one can related the measured strain to the angle between joints. The Bragg pitch deviates in accordance to the finger's flexion and extension movements.

A human hand has 14 joints to be monitor. Therefore, the same number of FBGs is required to be positioned at each joint. The FBGs' inherent multiplexing and self-referencing characteristics helps to reduce the system complexity as all the required sensors can be fitted in a single optical fiber (see the Figure 6).

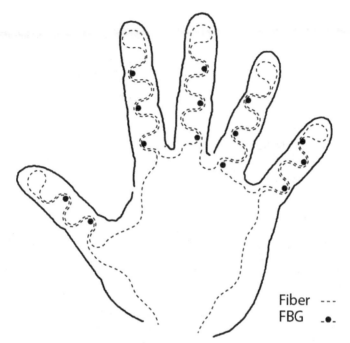

Fiber ---
FBG -•-

Figure 6. FBG sensor positioning proposal.

The nominal elongation of the finger joints, at the top or bottom face of the hand, is around 14 %, which is higher that the optical fiber's elongation range. Such limitation can be overcome in at least three methods by playing with the sensor positioning and/or optical fiber layout:

- The sensor may be placed on the side face of the joint. It is known that at a midline of a structure, the elongation while bending is null. A similar situation occurs on the finger joints, as one can see it as a bending load applied to the finger. Closer the FBG is positioned to the midline, lower is the strain that it will undergo. However, there is a trade-off associated to the movement sensitivity.
- The optical fiber can be coiled around the finger. This would create a spring effect on the optical fiber as the finger performs flexion and extension.
- Place the fiber in a curvilinear layout over the upper-face plane of the hand. This simulates the previous coil effects but on a two dimensional plane.

Based on the developed technique to integrate FBGs on flexible polymeric laminates, one could fabricate such laminate and use as a glove's upper face. On the upper face, the sensors are positively stretched, avoiding the wrinkles effect that occurs on the glove's lower face.

The laminate fabrication process ensures the correct positioning of the sensors and, at the same time, it provides protection to the optical fiber, enabling reliability and improving the wear out.

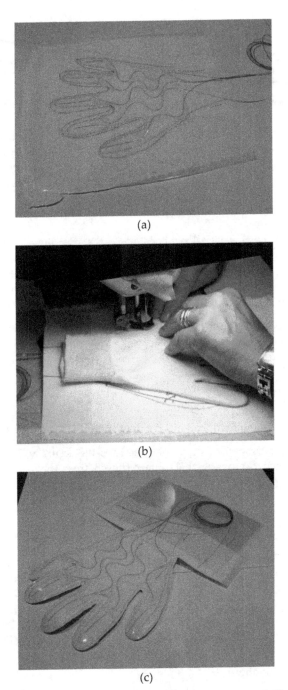

Figure 7. (a) Hand-shape polymeric foil with embedded FBGs; (b) Polymeric foil being sewed to a textile glove; (c) FBG-instrumented glove functional prototype.

3.4. Glove fabrication

Although the bare optical fiber commonly used for FBG only has an external diameter of 250 µm, which makes it perfect to be embedded in many structures, its fragility is still a major issue. The developed solution based on the integration of optical fiber and FBGs inside a thin and flexible polymeric substrate facilitates its use as a garment. It has been reported its flexibility, stretchability and capacity to keep the bond between the optical fiber and the substrate for signal transduction (Silva A. F. et al., 2009; Silva A. F. et al., 2010a). The degree of customization of the flexible substrate enables it to be manufactured with a hand shape. The approach was to replace the upper face of a glove by the polymeric substrate with the embedded FBGs in a single optical fiber (see the photographs in the Figure 7).

3.5. Performance assessment

The Figure 8 shows the raw signal obtained while the ring finger performs flexion and extension movements. The raw data shows the Bragg pitch deviation along the time the subject perform opening and closing hand movements. It is important to remark that the retrieved data is only related to the proximal interphalangeal crease of the ring finger. From the raw data, information can be processed namely, range of motion, strength and movement speed.

As the sensors are positioned on the glove's upper face, as the subject closes the hand, a positive strain occurs on the sensor site, resulting in a positive deviation of the Bragg pitch. As one opens the hand, the Bragg pitch decreases. For data processing, the null deviation occurs when the subject has his hand open, while the maximum deviation is set to the close hand state, driving the maximum strain.

Indirectly, the strength may be determined, since there is a correlation between the measured strain and the required load of 128 pm.N^{-1} (Silva A. F. et al., 2010b).

An important characteristic in this type of systems is the accuracy. For this system, such parameter is evaluated comparing the value retrieved from the system with the valued measured by a goniometer. A FBG-based system is able to present an almost true linear response - see the Figure 8(b) - with a maximum error of 2º in a 90º range.

Although the system mainly monitors the flexion-extension movement, one could experience that the acquired data is not influenced by the abduction-adduction, as it is constrained by the glove's structure itself.

Another key factor while developing the sensing glove is related to the Bragg wavelength inscription for each one of the 14 FBGs sensors. The reflected spectral component of each FBG uses around 0.3 nm of the available spectrum and requires a dynamic range of 1 nm for the finger movement. By considering the C-band optical range (1530-1560 nm), each FBG should be inscribed in 1.84 nm steps in order to fit all sensors in a single fiber.

3.6. Virtual hand movement

The therapy session not only is tedious for the therapist as it also monotonous to the patient. In order to improve the motivation surrounding the therapeutic session, a virtual reality environment may be set up based on the developed monitoring system. As the data that is retrieved from the sensors is made via personal computer, besides presenting the data exclusively for the therapist, the data can be used to create a tridimensional model of the hand that replicates what the patient is doing.

Furthermore, virtual interaction can be added, enabling game-like and personalized paced exercises to promote finger strength while keeping the subject motivated.

In the developed environment (see the Figure 9), it is possible to visualize the hand movement in real-time and provide at the same time information about the hand posture in terms of angles, strength and movement range.

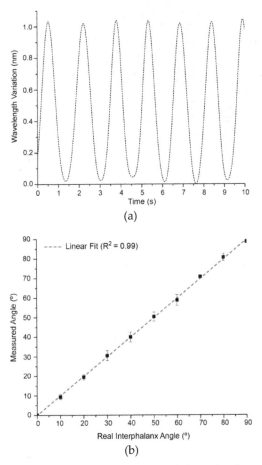

(a)

(b)

Figure 8. (a) Ring finger FBG sensor response for opening and closing hand movements; (b) System accuracy based on the comparison between real and measured angles.

A virtual monitoring system can be developed in LabView® environement with two purposes:

- Stimulate the patient to the therapy sessions by establishing game-like exercises. The hand movements are replicated in real time on the virtual environment that can be rotated and span.
- Provide information to the therapist regarding the angle at each joint, range of motion, movements speed and strength. It also provides a database of records that are helpful to evaluate the treatment evolution.

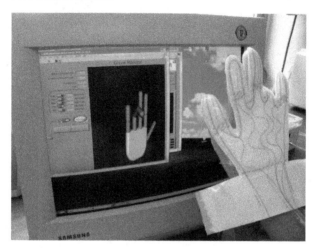

Figure 9. An example of the real-time monitor of the hand posture.

The monitoring hardware enables sampling frequencies from 32 Hz up to 2 kHz, ensuring a smooth visualization of the hand movement. The computational requirements are quite low (Intel Pentium IV processor with 512 MB of RAM), meaning that this systems does not require any special configuration, which is of great interest as it reduces the cost.

4. Cardio-respiratory frequency monitoring

The ability to monitor the vital signs of patients requiring medical assistance is a crucial issue (Fernandez R. et al., 2005), where the respiratory and the cardiac frequencies can be selected (among others) as presenting high interest (Evans D. et al., 2001). There are specific situations where the acquisition of these frequencies are important, e.g., with patients doing exams based on Magnetic Resonance Imaging (or MRI). However, this can be unsuitable, due to the potential occurrence of thermal or electrical burns associated with oximeter sensors and cables, temperature probes and MRI surface coils. These burns can be a result of inductions during MRI exams (Dempsey M. F. et al., 2001; Jones S. et al., 1996) or even during cardiothoracic surgeries (Wehrle G. et al., 2001). Therefore, the use of optical fibers can be an interesting solution for measuring the cardio-respiratory frequency. This statement is of particular importance because the optical fibers don't contain conductive parts therefore, this makes the optical fibers insensitive to external electromagnetic fields. In this section it will be proved the exequibility to deploy sensing/monitoring solutions for both the pulmonary and the cardiac frequency. Contrary to similar solutions found in the literature (Augousti A. T. et al., 2005; Davis C. et al., 1999), this section shows how it is possible to use a single optical fiber sensor and at the same time keeping it compatible with the healthcare environments.

4.1. Approach

There are few requirements that must be addressed: providing a simple sensing solution capable of measuring the cardio-pulmonary components with a single sensor and at the same time ensures their compatibility with different people. As illustrated in the Figure 10, the most suitable approach for complying with these requirements is providing a small and flexible structure able to readily be attached/unattached to/from the chest's site. Such approach enables their use by any person. A fixation mechanism must also be provided in order to ensure that the sensing element is able to follow the elongation of the chest wall due to respiratory and cardiac components. This flexible structure is designated as carrier material and was used with the intention to increase the strain sensitivity of the FBG sensors, as well as, to improve the adhesion.

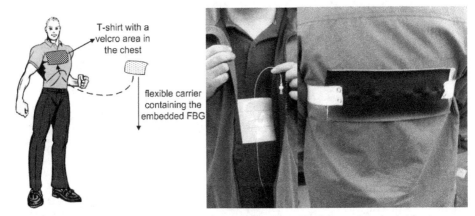

Figure 10. An illustration and two photographs with two views (the front and the backside views are respectively the left and right photographs) of the proposed approach showing the carriers containing the FBGs sensors.

4.2. Methodology

The Figure 11 shows a functional prototype that was tested on a group of few healthy subjects with ages between twenty and thirty years old. During the take of the measurements, the subjects were standing up and maintaining the full body to rest. It must be noted that the sensing foil was placed over the chest because this is the position of the human body where the effects of the heart beats are more significant.

The Figure 12 shows the block diagram of the complete FBG acquisition system. This filtering system was implemented for separating the respiratory and cardiac components from the acquired FBG signals. This system uses two band-pass filters, e.g., one is tuned in the 0.1-0.4 Hz range for allowing the measurement of the respiratory frequency. The pass-band of the second filter rejects all spectral components, except those in the range 0.5-1.3 Hz for retrieving the cardiac frequency. This second band-pass filter allows the discrimination of frequency around 1 Hz. This set with the frequencies of interest are obtained by cutting the respiratory components below 0.5 Hz and the high frequencies (mainly composed by high frequency noise) above 1.3 Hz. A software application was used to implement both pass-band filters in the digital domain using the bilinear technique (Losada R. A. et al., 2005) with a sampling frequency of 36 Hz.

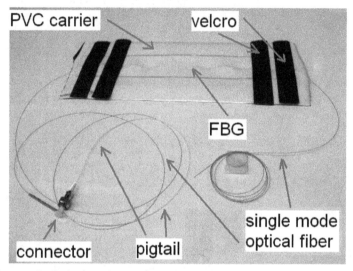

Figure 11. Photograph of a functional prototype (Silva A. F. et al., 2011a).

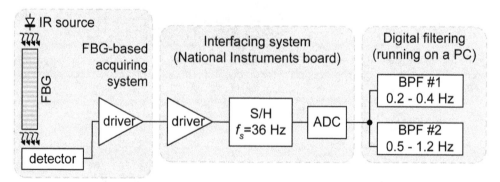

Figure 12. The block diagram of the FBG acquisition system (Silva A. F. et al., 2011a).

4.3. Experimental: Respiratory frequency

In the first trials, the subjects breathing naturally for evaluating the raw signal without any processing stage. As illustrated in the Figure 13, the external interferences do not appear to degrade the quality of the acquired signals. The small perturbations that were observed are mainly due to the transition between the inhale and the exhale stages.

The ability to establish a relation between the wavelength deviation to other quantities (e.g., the displacement and force) is one advantage of these sensors and their linear response to strain. As the FBG spectral signature deviates 8 nm per 1% of elongation (Silva A. F. et al., 2010b), it is possible to retrieve how much did the chest stretched. Consequently, the air volume that is inhaled and exhaled can also be estimated as well as the force that is being applied to breath. As the chest elongation can be retrieved from the Bragg pitch deviation, the volume of air inhaled or exhaled can be determined, as the Bragg sensors responds linear to strain (Silva A. F. et al., 2011c). A similar approach can be used to obtain the applied load to inhale, since there is also a linear relationship between the elongation and the necessary load (Silva A. F. et al., 2010a). The Figure 14 shows the corresponding frequency spectrums, which confirms the existence of main frequency peak between 0.1 and 0.4 Hz. In this figure is also possible to observe the group of high-frequency components superimposed on the normal respiratory signal that may be originated from involuntary body movements.

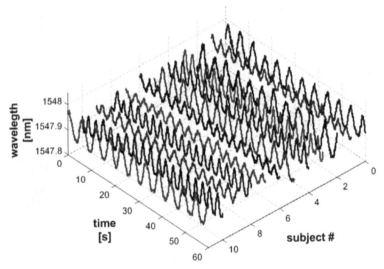

Figure 13. The sensor response to a normal breath (raw data) that was obtained for a group of twelve healthy subjects (#0 to #11).

These respiratory results were compared with reference signals acquired with the help of a commercial device (e.g., the Zephyr BioHarness) for validation purposes. The raw signals acquired from the FBGs were subjected to a band-pass filter in the 0.1-0.4 Hz range. The Figure 15 shows the signals of a single subject that have been acquired with the help of both the commercial device and the FBG sensor. It is possible to confirm quasi-identical behaviors along the time for the signals variations. However, a few differences may be due to the signal processing stage of the commercial device to which there was no access to.

Nevertheless, the same respiratory frequency (e.g., about 24 inhales per minute) was determined in both signals, and therefore, validating the measurements with FBGs.

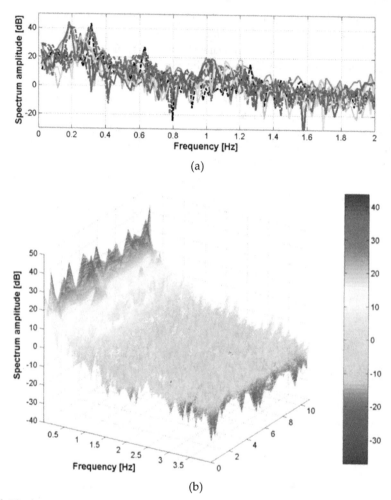

(a)

(b)

Figure 14. The frequency spectrum of normal breath also obtained for a group of twelve healthy subjects (#0 to #11). There top plot shows a superposition of all frequency spectrums for achieving a better visualization of the breading peaks. The bottom plot allows a better visualization of all frequency spectrums in the whole frequency range.

4.4. Experimental: Cardiac frequency

It exist a different test where the subjects are asked to do a deep inhale and a halt on its breath (once again, the Figure 16 shows an example for a single subject). At this point, a higher frequency response was obtained when compared with the respiratory frequency. The observation of these results led to the assumption that this behavior was related to the heart beat frequency. The signals showed in the Figure 17 correspond to the cardiac part and were retrieved with the help of the signal processing stage illustrated in the Figure 12. The respective frequency spectrums are illustrated in the Figure 18, where it is a clear the existence of a region with the location of the cardiac frequency peaks in the range 0.5-1.3 Hz.

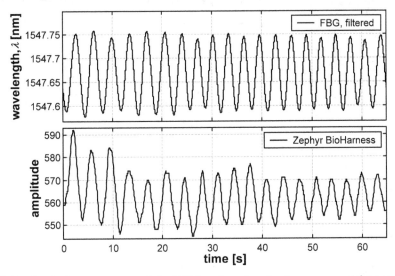

Figure 15. For the respiratory component: the FBG-based sensing structure response (plot on top) versus the Zephy BioHarness commercial response (plot on bottom).

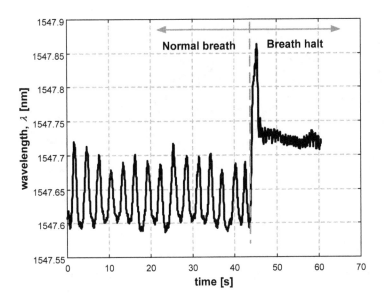

Figure 16. Sensor's response to a normal breath followed by a breath halt.

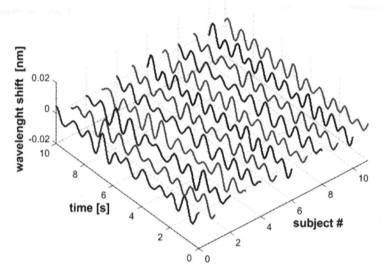

Figure 17. The cardiac frequency signals obtained by filtering the acquired raw data from the twelve healthy subjects (#0 to #11).

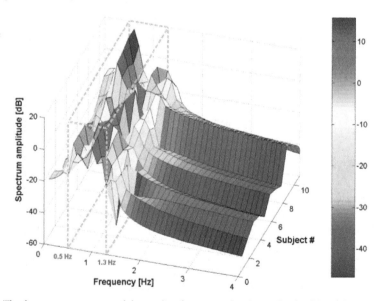

Figure 18. The frequency spectrum of the cardiac frequency for the twelve healthy subjects (#0 to #11).

The validation of the cardiac components was also done by comparing the obtained results with the commercial device previously used. The comparison between both systems is showed in the Figure 19. This test was also done for a single test subject because their cardiac frequency peaks are all located in the expected frequency range, e.g., between 0.5 Hz and 1.3 Hz. The FBG sensor and the flexible carrier in PVC present a similar behavior in comparison with the commercial system, e.g., 66 heartbeats per minute on both. Therefore, this approach makes possible to retrieve the information about the cardiac frequency. It must be noted that some of the lag between the both signals is a result from external factors. Such factor includes involuntary movements and irregularities in the breath.

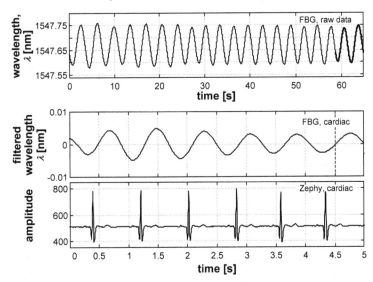

Figure 19. For the cardiac component retrieving: the FBG sensing structure raw data (top plot) and cardiac frequency (middle plot) and comparison with commercial system (bottom plot).

5. Epilogue

This chapter presented biomedical applications for acquisition systems based on FBGs. The absence of mechanical steps on sensor's fabrication results in the possibility to fabricate high sensitivity sensors with high reproducibility of their characteristics (Hill K. O. *et al.*, 1997). However, the most important features that made FBG-based systems a wide established technology were their electrically passive operation, electromagnetic interference immunity, compact size, self referencing capability, and more important, inherent multiplexing-ability, which enable a wide number of sensors in a single fiber as well as Bragg a single interrogation system (Wang Q. *et al.*, 2007). In conclusion, the FBGs are not restricted for the applications presented in this chapter.

Author details

Alexandre Ferreira da Silva
MIT Portugal Program, School of Engineering, University of Minho, Guimarães, Portugal

Rui Pedro Rocha, João Paulo Carmo and José Higino Correia
Department of Industrial Electronics, University of Minho, Guimarães, Portugal

Acknowledgement

This work was fully supported by the Algoritmi's Strategic Project UI 319-2011-2012, under the Portuguese Foundation for Science and Technology grant Pest C/EEI/UI0319/2011. Furthermore, the PhD student Rui Pedro Rocha was fully supported by the PhD scholarship with the reference SFRH/BD/33733/2009.

6. References

Afonso, J. A., Macedo, P., Silva, H. D., Correia, J. H., Rocha, L. A. (2010). "Design and Implementation of Multi-User Wireless Body Sensor Networks", *Journal on Advances in Networks and Services, IARIA Publisher,* 3(1&2): 62-74.

Alton, F., Baldey, L., Caplan, S., Morrissey, M.C. (1998). "A kinematic comparison of overground and treadmill walking", *Clinical Biomechanics,* 13(6): 434-440.

Anderson, D., Sidaway, B. (1994). "Coordination changes associated with practice of a soccer kick", *Research Quarterly Exercise and Sport,* 65(2): 93-99.

Antunes, P., Varum, H., and André, P. (2011). Uniaxial fiber Bragg grating accelerometer system with temperature and cross axis insensitivity", *Measurement,* 44(1): 1-5.

Arregui, F. J., Matias, I. R., Cooper, K. L., and Claus, R. O. (2002). Simultaneous measurement of humidity and temperature by combining a reflective intensity-based optical fiber sensor and a fiber bragg grating, *IEEE Sensors Journal,* 2(5): 482-487.

Augousti, A. T., Maletras, F. X., and Mason, J. (2005). Improved fibre optic respiratory monitoring using a figure-of-eight coil, *Physiological Measurement,* 26(5): 585-590.

Bao, H., Dong, X., Zhao, C., Shao, L. Y., Chan, C. C., and Shum, P. (2010). Temperature insensitive FBG tilt sensor with a large measurement range, *Optics Communications,* 283(6): 968-970.

Beard, P. C., Mills, T. N. (1996). "Extrinsic optical-fiber ultrasound sensor using a thin polymer film as a low-finesse Fabry-Perot interferometer", *Applied Optics,* 35(4): 663-675.

Bilro, L., Oliveira, J. G., Pinto, J. L., Nogueira, R. N. (2011). "A reliable low-cost wireless and wearable gait monitoring system based on a plastic optical fibre sensor", *Measurement Science and Technology,* 22(4): 1-7. Institute of Physics Paper 045801.

Boerkamp, M., Lamb, D. W., Lye, P. G. (2007, July). "Using an intrinsic, exposed core, optical fibre sensor to quantify chemical scale formation", *Journal of Physics: Conference Series,* 76(1): 1-7. Institute of Physics Paper 012016.

Carmo, J. P., Silva, A. F., Rocha, R. P., and Correia, J. H. (2012). Application of fiber Bragg gratings to wearable garments, *IEEE Sensors Journal,* 12(1): 261-266.

Cup, E. H., Pieterse, A. J., ten Broek-Pastoor, J. M., Munneke, M., van Engelen, B. G., Hendricks, H. T., van der Wilt, G. J., Oostendorp, R. A. (2007). "Exercise therapy and other types of physical therapy for patients with neuromuscular diseases: A systematic review", *Archives of Physical Medicine and Rehabilitation,* 88(11): 1452-1464.

Davis, C., Mazzolini, A., Mills, J., and Dargaville, P. (1999). A new sensor for monitoring chest wall motion during high-frequency oscillatory ventilation, *Medical Engineering & Physics,* 21(9): 619-623.

Dempsey, M. F., and Condon, B. (2001). Thermal injuries associated with MRI, *Clinical Radiology*, 56(6): 457-465.

Denselight Semiconductors. (2012). [on-line, 25th June 2012]: http://www.denselight.com/.

Dipietro, L., Sabatini, A. M., and Dario, P. (2003). "Evaluation of an instrumented glove for hand-movement acquisition", *Journal of rehabilitation research and development*, 40(2): 179-189.

Evans, D., Hodgkinson, B., and Berry, J. (2001). Vital signs in hospital patients: a systematic review, *International Journal of Nursing studies*, 38(6): 643-650.

Fahn, C.-S., and Sun, H. (2005). "Development of a data glove with reducing sensors based on magnetic induction", *IEEE Transactions on Industrial Electronics*, 52(2): 585-594.

Fender, A., MacPherson, W. N., Maier, R., Barton, J. S., George, D. S., Howden, R. I., Smith, G. W., Jones, B. McCulloch, S., Chen, X., Suo, R., Zhang, L., and Bennion, I. (2008). Two-axis temperature-insensitive accelerometer based on multicore fiber bragg gratings, *IEEE Sensors Journal*, 8(7): 1292-1298.

Fernandez, R., and Griffiths, R. (2005). A comparison of an evidence based regime with the standard protocol for monitoring postoperative observation: a randomised controlled trial, *The Australian journal of advanced nursing: a quarterly publication of the Royal Australian Nursing Federation*, 23(1): 15-21.

FiberSensing Sistemas Avançados de Monitorização, S. A. (2012). [on-line, 25th June 2012]: http://www.fibersensing.com/.

Godfrey, A., Conway, R., Meagher, D., and ÓLaighin, G. (2008). "Direct measurement of human movement by accelerometry", *Medical Engineering & Physics*, 30(10): 1364-1386.

Grattan, S. K. T., Taylor, S. E., Sun, T., Basheer, P. A. M., and Grattan, K. T. V. (2009). "In-situ cross-calibration of in-fiber bragg grating and electrical resistance strain gauges for structural monitoring using an extensometer", *IEEE Sensors Journal*, 9(11): 1355-1360.

Grillet, A., Kinet, D., Witt, J., Schukar, M., Krebber, K., Pirotte, F., and Depré, A. (2008). "Optical fiber sensors embedded into medical textiles for healthcare monitoring", *IEEE Sensors Journal*, 8(7): 1215-1222.

Gu, X., Guan, L., He, Y., Zhang, H. B., and Herman, P. R. (2006). High-strength fiber bragg gratings for a temperature-sensing array, *IEEE Sensors Journal*, 6(3): 668-671.

Hill, K. O., and Meltz, G. (1997). Fiber Bragg grating technology fundamentals and overview, *IEEE Journal of Lightwave Technology*, 15(8): 1263-1276.

Ibsen Photonics. (2012). [on-line, 25th June 2012]: http://www.ibsen.dk/.

Jones, S., Jaffe, W., and Alvi, R. (1996). Burns associated with electrocardiographic monitoring during magnetic resonance imaging. *Journal of the International Society for Burn Injuries*, 22(5): 420-421.

Kang, H. G., and Dingwell, J. B. (2008). "Separating the effects of age and walking speed on gait variability", *Gait & Posture*, 27(4): 572-577.

Kun, L., Inoue, Y., Shibata, K., and Enguo, C. (2011). "Ambulatory estimation of knee-joint kinematics in anatomical coordinate system using accelerometers and magnetometers", *IEEE Transactions on Biomedical Engineering*, 58(4): 435-442.

Lee, B. (2003). "Review of the present status of optical fiber sensors", *Optical Fiber Technology*, 9(2): 57-79.

Ling, H. Y., Lau, K. T., Cheng, L., and Jin, W. (2006). Viability of using an embedded FBG sensor in a composite structure for dynamic strain, *Measurement*, 39(4):328-334.

Lloyd-Jones, D., Adams, R. J., Brown, T. M., Carnethon, M., Dai, S., De Simone, G., Ferguson, T. B., et al. (2010). "Heart Disease and Stroke Statistics 2010 Update: a Report from the American Heart Association", *Circulation*, 121: 173.

Lorussi, F., Scilingo, E. P., Tesconi, A., Tognetti, A., and De Rossi, D. (n.d.). Wearable sensing garment for posture detection, rehabilitation and tele-medicine, Proceedings of the 4th International IEEE EMBS Special Topic Conference on Information Technology Applications in Biomedicine, 287-290.

Lorussi, F., Scilingo, E. P., Tesconi, M., Tognetti, A., and De Rossi, D. (2005). "Strain sensing fabric for hand posture and gesture monitoring", *IEEE transactions on information technology in biomedicine*, 9(3): 372-381.

Losada, R. A., and Pellisier, V. (2005). Designing IIR filters with a given 3-dB point, *IEEE Signal Processing Magazine*, 22(4):95-98

Mavrogiorgoua, P., Mergla, R., Tiggesa, P., Husseinia, J., Schrötera, A., Juckela, G., Zaudigb, M., and Hegerla, U. (2001). "Kinematic analysis of handwriting movements in patients with obsessive-compulsive disorder", *Journal of Neurology, Neurosurgery & Psychiatry*, 70(5): 605-612.

Moustakidis, S. P., Theocharis, J. B., and Giakas, G. (2010). "A fuzzy decision tree-based SVM classifier for assessing osteoarthritis severity using ground reaction force measurements", *Medical Engineering & Physics*, 32(10): 1145-1160.

Nishiyama, M., and Watanabe, K. (2009). "Wearable sensing glove with embedded hetero-core fiber-optic nerves for unconstrained hand motion capture", *IEEE Transactions on Instrumentation and Measurement*, 58(12): 3995-4000.

Oplink Communications Inc. (2012). [on-line, 25th June 2012]: http://www.oplink.com/.

Orr, P., and Niewczas, P. (2010). An optical fiber system design enabling simultaneous point measurement of magnetic field strength and temperature using low-birefringence FBGs, *Sensors and Actuators A: Physical Sensors*, 163(1): 68-74.

Parker, T. M., Osternig, L. R., van Donkelaar, P., and Chou, L.-S. (2008). "Balance control during gait in athletes and non-athletes following concussion", *Medical Engineering & Physics*, 30(8): 959-967.

Patterson, K. K., Nadkarni, N. K., Black, S. E., and McIlroy, W. E. (2012). "Gait symmetry and velocity differ in their relationship to age", *Gait & Posture*, 35(4): 590-594.

Peng, B. J., Zhao, Y., Yang, J., and Zhao, M. (2005). Pressure sensor based on a free elastic cylinder and birefringence effect on an FBG with temperature-compensation, *Measurement*, 38(2): 176-180.

Peng, B.-J., Zhao, Y., Zhao, Y., and Yang, Y. (2006). Tilt sensor with FBG technology and matched FBG demodulating method, *IEEE Sensors Journal*, 6(1): 63-66.

Perng, J. K., Fisher, B., Hollar, S., and Pister, K. S. J. (n.d.). Acceleration sensing glove (ASG), Proceedings of the 3rd International Symposium on Wearable Computers.

Rajan, G., Callaghan, D., Semenova, Y., McGrath, M., Coyle, E., and Farell, G. (2010). A fiber bragg grating-based all-fiber sensing system for telerobotic cutting applications, *IEEE Sensors Journal*, 10(12): 1913-1919.

Ren, L., Jones, R. K., and Howard, D. (2008). "Whole body inverse dynamics over a complete gait cycle based only on measured kinematics", *Journal of Biomechanics*, 41(12): 2750-2759.

Riley, P. O., Paolini, G., Croce, U., Paylo, K. W., and Kerrigan, D. C. (2007). "A kinematic and kinetic comparison of overground and treadmill walking in healthy subjects", *Gait & Posture,* 26(1): 17-24.

Rocha, R. P., Silva, A. F., Carmo, J. P., and Correia, J. H. (2011). FBG Sensor for measuring and recording the knee joint movement during gait, Proceedings of the 33rd Annual International Conference of the IEEE Engineering in Medicine and Biology Society (EMBC '11), Boston, Massachusetts, USA.

Scilingo, E. P., Lorussi, F., Mazzoldi, a., and De Rossi, D. (2003). "Strain-sensing fabrics for wearable kinaesthetic-like systems", *IEEE Sensors Journal*, 3(4): 460-467.

Silva, A F, Goncalves, F., Ferreira, L. A., Araújo, F. M., Dias, N. S., Carmo, J. P., Mendes, P. M., and Correia, J. H. (2009). Manufacturing technology for flexible optical sensing foils, Proceedings of the 35th Annual Conference of IEEE Industrial Electronics (IECON 2009), Oporto, Portugal.

Silva, A. F., Gonçalves, F., Ferreira, L. A., Araújo, F. M., Mendes, P. M., and Correia, J. H. (2010a). "Fiber Bragg grating sensors integrated in polymeric foils", *Materials Science Forum*, 636-637: 1548-1554.

Silva, A. F., Goncalves, A. F., Ferreira, L. A., Araujo, F. M., Mendes, P. M., and Correia, J. H. (2010b). PVC smart sensing foil for advanced strain measurements, *IEEE Sensors Journal*, 10(6): 1149-1155.

Silva, A. F., Carmo, J. P., Mendes, P. M., and Correia, J. H. (2011a). Simultaneous cardiac and respiratory frequency measurement based on a single fiber Bragg grating sensor, *Measurement Science and Technology*, 22(7): 1-5. Institute of Physics Paper 075801.

Silva, A. F., Goncalves, A. F., Mendes, P. M., and Correia, J. H. (2011b), FBG sensing glove for monitoring hand posture, *IEEE Sensors Journal*, 11(10): 2442-2448.

Silva, A. F., Goncalves, A. F., Ferreira, L. A., Araujo, F. M., Mendes, P. M., and Correia, J. H. (2011c). A Smart skin PVC foil based on FBG sensors for monitoring strain and temperature, *IEEE Transactions on Industrial Electronics*, 58(7): 2728-2735.

Silva, A. F., Gonçalves, A. F., Mendes, P. M., and Correia, J. H. (2012). PVC formulation study for the manufacturing of a skin smart structure based in optical fiber elements, *Journal Polymers for Advanced Technologies, Wiley Publisher*, 23(2), 220-227.

Thorlabs GmbH. (2012). [on-line, 25th June 2012]: http://www.thorlabs.de/.

Tognetti, A., Carbonaro, N., Zupone, G., and De Rossi, D. (2006). Characterization of a novel data glove based on textile integrated sensors, Proceedings of the Annual International Conference of the IEEE Engineering in Medicine and Biology Society.

Trew, M., and Everett, T. (2005). Human movement: an introductory text, 5th edition. Elsevier Limited.

Turcot, K., Aissaoui, R., Boivin, K., Pelletier, M., Hagemeister, N., and Guise, D. (2008). "New accelerometric method to discriminate between asymptomatic subjects and patients with medial knee osteoarthritis during 3-D gait", *IEEE Transactions on Biomedical Engineering*, 55(4): 1415-1422.

Vancampfort, D., M. Probst, Skjaerven, L. H., Catalán-Matamoros, D., Lundvik-Gyllensten, A., Gómez-Conesa, A., Ijntema, R., and De Hert, M. (2012). "Systematic review of the benefits of physical therapy within a multidisciplinary care approach for people with schizophrenia", *Journal of the American Physical Therapy Association*.

von Porat, A., Henriksson, M., Holmström, E., and Roos, E. M. (2007). "Knee kinematics and kinetics in former soccer players with a 16-year-old ACL injury - the effects of twelve weeks of knee-specific training", *BMC Musculoskeletal Disorders*, 8(35).

Wang, Q., Rajan, G., Wang, P., and Farrell, G. (2007). Macrobending fiber loss filter, ratiometricwavelength measurement and application, *Measurement Science Technology*, 18(10): 3082-3088.

Wehrle, G., Nohama, P., Kalinowski, H. J., Torres, P. I., and Valente, L. C. G. (2001). A fibre optic Bragg grating strain sensor for monitoring ventilatory movements, *Measurement Science and Technology*, 12(7): 805-809.

Wei, C.-L., Lai, C.-C., Liu, S.-Y., Chung, W., Ho, T., Ho, S., McCusker, A., Kam, J., and Lee, K. (2010). "A fiber bragg grating sensor system for train axle counting". *IEEE Sensors Journal*, 10(12): 1905-1912.

Wu, Y., and Shi, L. (2011). "Analysis of altered gait cycle duration in amyotrophic lateral sclerosis based on nonparametric probability density function estimation", *Medical Engineering & Physics*, 33(3): 347-355.

Xie, F., Chen, Z., and Ren, J. (2009). Stabilisation of an optical fiber Michelson interferometer measurement system using a simple feedback circuit, *Measurement*, 42(9): 1335-1340.

Yamamoto, Y. (2004). "An alternative approach to the acquisition of a complex motor skill: Multiple movement training on tennis strokes", *International Journal of Sport and Health Science*, 2: 169-179.

Yang, X. J., Hill, K., Moore, K., Williams, S., Dowson, L., Borschmann, K., Simpson, J. A., and Dharmage, S. C. (2012). "Effectiveness of a targeted exercise intervention in reversing older people's mild balance dysfunction: a randomized controlled trial", *Journal of the American Physical Therapy Association*.

Yao, S.-K., and Asawa, C. (2003). "Fiber Optical Intensity Sensors", *IEEE Journal on Selected Areas in Communications*, 1(3): 562-575.

Yavuzer, G., Öken, Ö., Elhan, A., and Stam, H. J. (2008). "Repeatability of lower limb three-dimensional kinematics in patients with stroke", *Gait & Posture*, 27(1): 31-35.

Yeo, T. L., Sun, T., Grattan, T. K. V., Parry, D., Lade, R., and Powell, B. D. (2005). Polymer-coated fiber Bragg grating for relative humidity sensing, *IEEE Sensors Journal*, 5(5): 1082-1089.

Zhang, W., Li, F., and Liu, Y. (2009). FBG pressure sensor based on the double shell cylinder with temperature compensation, *Measurement*, 42(3): 408-411.

Zhao, Y., Zhao, Y., and Zhao, M. (2005). Novel force sensor based on a couple of fiber Bragg gratings, *Measurement*, 38(1): 30-33.

Chapter 7

Application of Fiber Bragg Grating Sensors in Power Industry

Regina C. S. B. Allil, Marcelo M. Werneck,
Bessie A. Ribeiro and Fábio V. B. de Nazaré

Additional information is available at the end of the chapter

1. Introduction

1.1. Application of a Fiber Bragg Grating temperature system in a grid-connected hydrogenerator

The main control parameters in any hydro-electric plant (HEP) or substation is, of course, the current and voltage. The third parameter is the temperature, which is normally a consequence of the current and must be kept under close observation because rises above 100°C may accelerate the aging of insulating materials and conductors or even destroy them [1].

With the idea of decreasing the number of copper wires and consequently decreasing installation and operation costs, we designed a fiber optic temperature sensor for application in large generators. The objective of the system is to cover all temperature monitoring needs of an HEP that would also overcome some of the disadvantages presented by the conventional RTD (resistive temperature detector) network.

This session describes the research, development and operation of an FBG temperature sensor array for hydro-electric generators. The optical fiber sensors system measured temperature at different points of an operational 42.5-MW generator at full power. The results showed feasibility and usefulness of the optical fiber system in power equipment.

1.2. Installation of sensors in the hydroelectric generator

The HEP chosen for this experiment was the 216 MW UHE-Samuel, in the western city of Porto Velho, Brazil. It is located on the Jamari River, a tributary of the Madeira River, which in turn, is one of the major tributaries of the Amazon River. Before being installed, the FBGs were calibrated in the laboratory to find their sensitivities [2].

The machine, which operates at a temperature of around 95°C at full load, needs 24 hours to drop its temperature to about 45°C to make it possible to enter the stator hall to install the optical sensors.

An FBG presents a very small time constant due to its small mass. To protect this sensor and not deteriorate such a valuable parameter, the sensor was loosely inserted inside a thin 10-cm U-form copper tubing to allow good heat transfer between the cooling air and the optical fiber, as shown in Figure 1. The copper tubing which also protects the sensor against strain comes out of and reenters an IP65 polymeric enclosure.

A conventional optical fiber cable connected all six boxes which were installed around the stator winding behind each radiator of the generator. The optical cable was then placed within the existing cable trays along with the other electric cables extending all the way up from the generator to the HEP control room where the optical interrogator and an industrial PC were installed.

The optical interrogation setup consists of a broad band optical source that illuminates all FBGs in the array. The return signal of each FBG is detected by the optical interrogator (Spectral Eye 400-FOS&S) that identifies the center wavelength of each FBG reflection pulse. Using the calibration parameters of each FBG, the optical interrogator calculates and stores the temperatures and communicates with an industrial PC via RS-232 interface.

The PC publishes all data on the company's Intranet and become available to the HEP central software control. Figure 1 shows the box containing one sensor installed inside the stator and Figure 2 shows a block diagram of the generator in detail with the optical cable connecting the six sensors to the interrogation system.

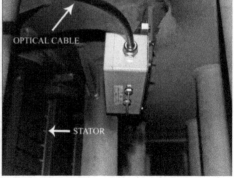

Figure 1. Box containing the optical fiber splices with the FBG inside the U-shape copper tubing (left) and installed inside the generator (right) (adapted from [2]).

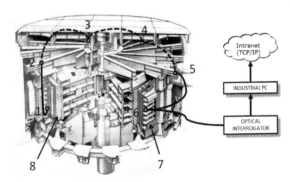

Figure 2. Depiction of a cross-section of the hydrogenerators showing the generator in detail with sensors connected to the interrogation system. 1-6) FBG sensors, 7) Radiator, 8) Stator (adapted from [2]).

1.3. Results

Immediately after the installation, the system started monitoring the temperatures, producing the graph shown in Figure 3. We can observe all signals superimposed at about 33°C.

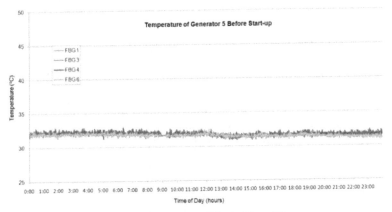

Figure 3. Temperature of generator 5 before start-up(adapted from [2]).

As part of the test procedures, the machine was started-up and shut down several times. The graph in Figure 4 shows the evolution of the temperature during the last start-up test of the generator. Notice that, in contrast to Figure 4, the temperatures of the radiators were not the same before the start-up. This is because previously the machine was working with different temperatures around the stator, which is normal, as we will see later. At 9 am the turbine was opened to the dam and the machine started-up. The temperature at FBG 3 rose from around 35°C to 85°C while the turbine accelerated to 90 rpm until in phase with the 60 Hz grid frequency. Then, at 6 pm the generator was switched to the national grid and the temperature rose again to 95°C, stabilizing thereafter.

Figure 5 shows the temperature of the generator when operating normally. At this time the generator was producing 22 MW with an average water flow of 82 m³/s.

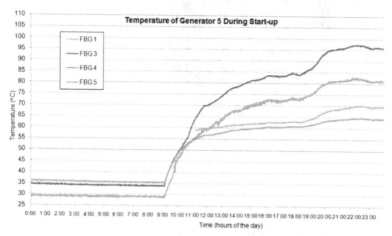

Figure 4. Temperature evolution of generator 5 during start-up(adapted from [2]).

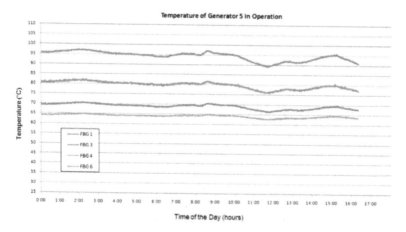

Figure 5. Temperature of generator 5 in operation (adapted from [2]).

It is worth pointing out that, in Figure 5 it is possible to see that, even in a steady state, the generator temperatures vary with time, with all temperatures following the same pattern. This is how the generator responds to the energy demands by the load.

As the final step of this project we needed to be sure that the temperature data monitored by our system was sufficiently accurate, after two years of operation inside the hydrogenerator. To accomplish this, we installed thermo-pair sensors which were physically attached to the FBGs copper tubing so as to compare conventional sensors with FBGs ones.

Thus we ended up with two sets of data for each radiator, one from our system itself and another one from the conventional thermo-pair datalog. Figure 6 shows the temperature evolution of one such measurements in which it is possible to see that both signals follow a similar pattern of behavior when the machine starts.

We can also observe that the thermo-pair sensor, being electronic and therefore subject to electromagnetic interference, produces a noisy output signal, differently from the FBG, which presents a smoother and asymptotic one, as expected.

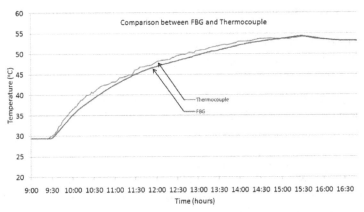

Figure 6. Comparison between an FBG sensor and a thermo-pair sensor as the machine starts (adapted from [2]).

1.4. Conclusions

This work describes the world's first application, test and operation of an FBG temperature sensor array inside a fully operational and connected-to-the-grid hydro-electric power generator.

A study of the sensing characteristics showed high repeatability, a maximum error of approximately 0.004°C (better than the calibrating thermometer) and linearity better than 99.9% for the five calibrated FBGs. These errors are related to the uncertainty of the interrogation system (±1 pm) together with an FBG average sensitivity of about 13 pm/°C. Therefore, 1 pm in error means a temperature uncertainty of about 0.08°C. Since uncertainties below one degree centigrade are quite acceptable in the electric power industry, the calibration experiment demonstrated that the FBG system is appropriated for temperature measurements.

With this system in operation, a huge amount of installation and maintenance costs could be avoided with the replacement of many kilometers of electric wires by a few optical cables.

As a final conclusion, the system was capable of reliably and accurately measuring and monitoring temperatures inside the generator, even considering the harsh environment of the stator.

2. DC high voltage measurement based in an FBG-PZT sensor

2.1. Introduction

Electric power facilities, such as substations, rely on two basic instruments for their functionality and protection: the voltage transformer (VT) and the current transformer (CT) for measuring and controlling voltage and current, respectively. Those equipment are reliable for over-voltage and over-current protection, allow 0.2% revenue metering accuracy and their behaviour is well known both under normal and abnormal conditions. Nevertheless these equipments are made entirely of copper, ceramic and iron with all empty spaces filled with oil, which are weighty materials, producing bulky, heavy and clumsy equipment. On top of that, they tend to explode without prior warning, resulting in the potential destruction of nearby equipment by pieces of sharp ceramics and furthermore putting the substation personnel at risk.

Optical voltage transducers offer many improvements over traditional inductive and capacitive voltage transformers. These advantages include linear performance and wider dynamic range, not to mention, lighter weight, smaller size, and improved safety. Since optical fibers carry the measurements as light signal to and from the sensor head, workers and all substation control equipment are electrically isolated from the high voltage environment. Due to their small footprint and light weight, these pieces of equipment allow the maintenance of emergency mobile substations that are deployed on site by trucks and start working in a matter of a few hours. Optical CTs and VTs for power systems have been used in the last decades and are commercially available today from a few companies [3].

On the other hand, both Pockels and Faraday effects have drawbacks. The first one is that both Faraday and Pockels system components are not stable with temperature and stress, demanding the use of complicated compensation techniques [4] or specially designed optical fibers [5] in order to become reliable. Additionally, Pockels cells are made of bulk crystals, e.g. lithiun niobate, demanding an open optics approach with lenses and polarizing filters which become an unstable and difficult-to-align system. The main drawback, however, is the high cost of this still new technology, not only for acquisition but also for maintenance, demanding specialty skills uncommonly available among company personnel.

From the demands of this work and the observations above, came the motivation for developing a practical VT system with two basic characteristics: a) intrinsic fiber optic sensors avoiding open optics and b) using a well-known and proven form of technology that would lead to a competitive price system as compared to conventional inductive VTs.

We have already designed and tested a hybrid electro-optical low cost monitoring system for the 13.8-kV distribution line [6], but nowadays the best technology for attending to the above demands is the fiber Bragg grating (FBG) that is relatively easy to deal with, reliable and very sensitive to strain. This section relates the development of a high voltage measuring system aimed to be used as the core of a 13.8-kV-class VT for the electric power industry application using as a sensing element a PZT crystal as voltage transducer and an FBG as strain measuring sensor.

2.2. Materials and methods

This section will develop the relationship between the resulted Bragg wavelength displacement when the optical fiber is submitted to a force applied by the PZT when the latter is subjected to an electric field.

As a first approach, the FBG was wound around a cylindrical PZT tube in a constant temperature environment as shown in the picture of Figure7.

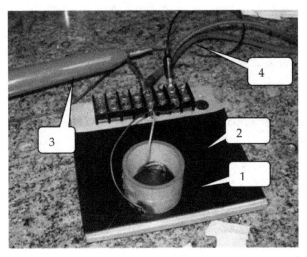

Figure 7. The optical fiber with a FBG (1) wound around the PZT crystal (2). Also shown are the high voltage probe (3) and the high voltage power supply leads (4).

When an electric field is applied to the PZT electrodes, the tube experiences both a circumference contraction (d_{31} is negative) and an increase in the wall thickness (d_{33} is positive). We can apply the following equation for calculating the net strain considering both effects.

For wall thickness displacement, we have:

$$\frac{\Delta w}{w} = d_{ij}E \tag{1}$$

where d_{ij} is the piezoelectric displacement coefficient (in meters per volt) with I being a designation for the direction crystal polarization and j being the Cartesian coordinate. The direction of polarization (axis 3) is established by a strong electrical field applied between the electrodes during the poling process.

Thus,

$$\Delta w = d_{33}Ew \tag{2}$$

But since the electrical field is

$$E = \frac{V_{in}}{w} \tag{3}$$

then we have:

$$\Delta w = d_{33} V_{in}. \tag{4}$$

Here V_{in} is the applied voltage and w is the wall thickness.

The cylinder wall thickness displacement will account for a variation of its diameter and consequently the length of the tube circumference, L, around which the fiber was wound.

Since $L=2\pi R$, we have $\Delta L=2\pi \Delta R$. Also, if the wall thickness is w, then, when the thickness increases by Δw, the radius R will increase by $\Delta w/2$, since half of the increase was in an inwards direction and does not account for ΔR. Then, we have:

$$\Delta L = 2\pi \frac{\Delta w}{2} = \pi \Delta w. \tag{5}$$

Substituting (4) into (5), we have the following fiber strain as a consequence of the wall thickness displacement:

$$\Delta L = \pi d_{33} V_{in} \tag{6}$$

But, as a result of the applied voltage, we also have a displacement of the circumference of the PZT tube. The relative increase of the circumference is, according to (2):

$$\frac{\Delta L}{L} = d_{31} E \tag{7}$$

or

$$\Delta L = 2\pi R d_{31} E \tag{8}$$

But, as $E=V_{in}/w$, we have:

$$\Delta L = 2\pi R d_{31} \frac{V_{in}}{w} \tag{9}$$

Finally, we add equations [6] and [9] to obtain the net effect:

$$\Delta L = V_{in} \left(\pi d_{33} + 2\pi R \frac{d_{31}}{w} \right) \tag{10}$$

or

$$\frac{\Delta L}{L} = V_{in} \left(\frac{d_{33}}{2R} + \frac{d_{31}}{w} \right) \tag{11}$$

For the PZT tube we have the following constants:

$d_{31}=-122$ pm/V; $d_{33}=300$ pm/V; $R=22.56\times10^{-3}$ m and $w=3.3\times10^{-3}$ m.

Substituting the above constants in (11) we have

$$\frac{\Delta L}{L} = -30.32\times10^{-9} V_{in}. \tag{12}$$

From Eq. (5) in [7] in a constant temperature environment and substituting the constants we have:

$$\frac{\Delta\lambda_B}{\lambda_B} = 0.78\frac{\Delta L}{L} \tag{13}$$

Substituting (12) in (13) we finally have:

$$\frac{\Delta\lambda_B}{\lambda_B} = -23.65 \times 10^{-9} V_{in} \tag{14}$$

Now, considering the central Bragg wavelength at rest, λ_B=1558.024 nm, we get to the following sensitivity for our system:

$$\frac{\Delta\lambda_B}{\Delta V} = -36.85 \times 10^{-3} \text{pm/V} \tag{15}$$

which means a Bragg wavelength shift of 36.85 pm per each 1000 V applied to the PZT.

In the next step, in order to increase the sensitivity, another approach was tested which is shown schematically in Figure 8. The PZT crystal was installed between the two levers and the fiber, with the FBG bonded on the levers' tips.

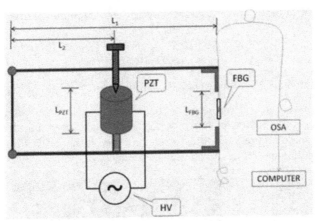

Figure 8. The setup with a variable-gain mechanical amplifier.

This setup consists of a mechanical amplifier with the possibility of varying the gain by changing the position of the tension screw along the upper lever.

The total gain G is, according to the parameters shown in Figure 8:

$$G = \frac{L_1}{L_2} \times \frac{L_{PZT}}{L_{FBG}} \tag{16}$$

This approach is based on the longitudinal mode strain of the PZT tube, that is, its length displacement. The position of the tension adjustment screw in the upper lever combined with the length of the PZT tube and the length of the optical fiber provided a mechanical strain amplification G=1.98.

We now have:

$$\frac{\Delta h}{h} = d_{32} \frac{V_{in}}{w} \tag{17}$$

Where $\frac{\Delta h}{h}$ is the relative change in height of the PZT tube, h is the height of the tube, w is the wall thickness and d_{32} is the piezoelectric coefficient in the longitudinal direction. These parameters have the following values: $h = 35$ *mm*; $w=3.3x10^{-3}$ *m* and $d_{32}= -122$ pm/V.

Substituting the above constants in (17) we get:

$$\frac{\Delta h}{h} = -36.97x10^{-9}V_{in}.$$

But since we have a multiplication factor of 1.98, the strain experienced by the fiber will be

$$\frac{\Delta L}{L} = -73.20x10^{-9}V_{in}. \tag{18}$$

Now substituting (18) in (13) we have:

$$\frac{\Delta \lambda_B}{\lambda_B} = -57.10x10^{-9}V_{in} \tag{19}$$

Considering a central Bragg wavelength at rest, $\lambda_B=1560.025$ nm, we arrive at the following sensitivity:

$$\frac{\Delta \lambda_B}{\Delta V} = -89.07x10^{-3}pm/V \tag{20}$$

which means a shift of 89.07 pm for each 1000 V applied to the PZT.

Figure 9 shows the complete setup with the Bragg meter and the high voltage power supply and Figure 10 shows a picture of the tube installed in the mechanical amplifier whose diagram was shown in Figure 7. The Bragg meter we used is the FS 2200 from Fiber Sensing, presenting a resolution of 1.0 pm and absolute accuracy of 2.0 pm.

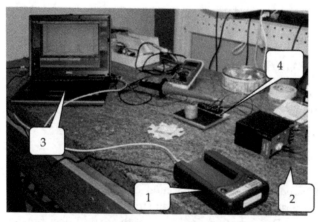

Figure 9. The complete setup showing the OSA (1), the high voltage power supply (2), the FBG spectrum (3) and the PZT crystal with the fiber wound around it (4).

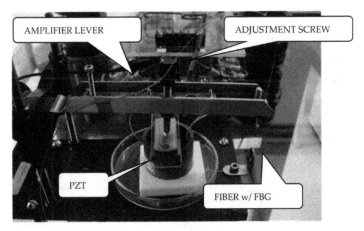

Figure 10. The PZT ceramic on the mechanical amplifier.

2.3. Results and discussions

For the first experiments, only DC voltages were applied to the PZT since the OSA we used to demodulate the FBG signal is too slow to respond to the 60Hz line frequency. The electro-optic setup was composed of an ASE light source, a 6 kV-DC variable power supply applied to the PZT, optical fibers inscribed with FBG and an OSA to trace the FBG's return signal and detect the center wavelength displacement.

In the first experiment the fiber was wound and glued around the PZT tube and for the second experiment the tube was fixed to the mechanical amplifier shown in Figure 10. This setup consists of a mechanical amplifier with the possibility of varying the gain by changing the position of the screw along the upper lever.

By applying a DC voltage to the PZT and recording the respective Bragg shift, it was possible to plot the graphs shown in Figure 11. The lower graph is the wavelength shift of the first experiment and the upper graph is the response of the second experiment. In this last experiment we used insulating oil in order to increase the applied voltage. The correlation coefficients (R^2) were 0.9936 and 0.9990 for the first and second experiment, respectively, which showed very good correlations and repeatability of the results in all measurements.

In the first experiment (fiber wound around PZT tube) the results obtained were shown by (15), or 36.85 pm per each 1000 V applied to the PZT. In this case we would have an inaccuracy of 13.6 V or about 1% of full scale. Obviously, this would not be satisfactory for a 0.2-class instrument transformer in revenue metering application.

In the second experiment (with mechanical amplifier) we obtained a wavelength shift of 89.07 pm for each 1000 V applied to the PZT (Equation (20). Notice that in Figure 11 it is clear that there is a good relationship between wavelength and the applied voltage with a correlation coefficient of 0.9990. So the inaccuracy of this measurement could only come

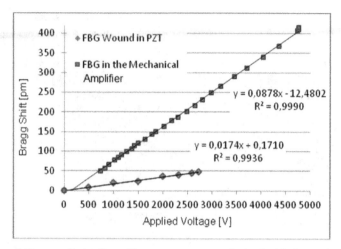

Figure 11. Bragg shift vs. applied voltage. The upper trace was obtained with the mechanical amplifier; the lower trace was obtained with the fiber wound around the PZT.

from the wavelength measurement by the Bragg Meter. Then, an inaccuracy in wavelength of 2.0 pm (the Bragg Meter's uncertainty) will lead to an inaccuracy in voltage of 22.45 V. In the TC's line voltage of 13.8 kV this error is 0.16%, meeting therefore the 0.2-Class of instrument transformer, IEC 60044-5 [8]. Since an increase in the gain (Eq. 16) leads to a larger sensitivity (Eq. 20) we can adjust the length L_2 in the amplifier and consequently decrease the error to a smaller value. Another alternative is to decrease the length of the FBG, L_{FBG} (Eq. 16) because this will also increase the gain.

2.4. Conclusions

The main conclusion of this work is that this setup can be used as a core of a practical 13.8 kV-Class VT if the issues mentioned above can be taken care off. Also, the difference from the maximum allowed voltage to be applied to PZT and the 13.8 kV expected to be measured can be easily circumvented by a capacitive divider. The divider will also be used when this system will be scalable to the 500kV-Class VT. These are preliminary results and a more appropriate setup is under development which will use the twin grating technique in order to demodulate AC voltages and a PZT disk stack to increase the longitudinal displacement and thus the fiber strain, resulting in an improved level of accuracy.

3. Measuring AC high voltage

3.1. Principle of operation of the FBG-PZT sensor

The experimental setup of the FBG-PZT sensor system is shown in Figure 12. The sensor was built by using a ceramic stack with ten 4-mm-thick PZT-4 rings, separated by 0.2-mm thick copper electrodes where the contacts were fixed. The proposed measurement system operates as follows: a voltage is applied in a combined PZT and FBG sensor by using a high voltage power supply. This voltage acts on the PZT ceramic causing a mechanical deformation that is transmitted to an attached FBG used as interrogation system. Hence, the spectrum of the reflected light from the FBG is detected and demodulated to obtain a signal proportional to the applied voltage.

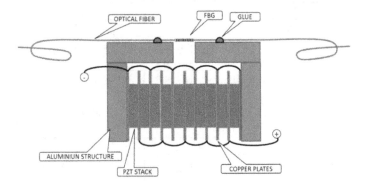

Figure 12. The mechanical setup comprised of ten ceramic disks in a stack. The polarization is made in such a way that all disks receive the same voltage [7].

The FBG with central wavelength of 1536.18 nm was stretched to 1538.48 nm before being bonded to the aluminum structure to allow measurements in both directions, since the PZT experiences positive and negative displacements when subjected to an AC voltage. For an improved isolation and to avoid sparks under higher voltages, the entire assembly was immersed in insulating oil.

For demodulating an FBG at DC operation one can use a commercial FBG interrogator as we can see in several low frequency applications. However, this equipment is limited to a few hertz, being incapable to respond to AC line frequency. We aimed the development of power line VT with an appropriated accuracy so as to attend the IEC [8] and also addressed to a study of temperature drift since temperature also displaces the Bragg wavelength, which is the measurement parameter. To calculate the sensitivity of the FBG-PZT transducer that is the relationship between the Bragg wavelength displacement and the electric filed applied to the PZT we start from:

$$\frac{\Delta w}{w} = d_{33}E \tag{21}$$

Where w is the PZT disk thickness, $E=V/w$ and V is the applied voltage. Now, since the FBG is bonded to the side of the stack, it will experience the same strain as the PZT and we can combine Eq.(5) in [7] with (21) yielding:

$$\frac{\Delta\lambda_B}{\lambda_B} = (1 - \rho_e)d_{33}\frac{V}{w} + (\alpha + \eta)\Delta T \tag{22}$$

and for a constant temperature environment, we have:

$$\Delta\lambda_B = \lambda_B(1 - \rho_e)d_{33}\frac{V}{w} \tag{23}$$

Table 1 summarizes the parameters and constants for FBG and the PZT ceramic used in this work.

FBG	
Physical property	**Value**
Bragg wavelength (25°C)	$\lambda_B = 1538.48$ nm
Photo-elastic coefficient	$\rho_e = 0.22$
Coefficient of thermal expansion	$\alpha = 0.55 \times 10^{-6}/°C$
Thermo-optic coefficient (dn/dT)	$\eta = 8.6 \times 10^{-6}/°C$
Length of fiber w/ FBG	$L = 28$ mm
PZT	
Physical property	**Value**
PZT type	PZT4
Ceramic shape	Ring
Piezoelectric strain constant	$d_{33} = 300$ pm/V
Thickness of ceramic	$w = 4$ mm
Maximum direct field strength	1-2 kV/mm
Maximum reverse field strength	350-500 V/mm
Curie Temperature	$T_c = 325°C$
Number of elements in stack	$n=10$

Table 1. PZT and FBG Parameters [7]

Substituting the PZT constants of Table I in (23) we have the following sensitivity:

$$\frac{\Delta\lambda_B}{\lambda_B} = 90.0 \text{ pm/kV} \tag{24}$$

which means a Bragg wavelength shift of 90 pm per each 1,000 V applied to the PZT. If we want to comply with the IEC 60044-5 [8] when measuring 13.8 kV, a conventional class for distribution line, we must attain 0.2% accuracy and so be capable to distinguish accurately at least 27.6 V when measuring the applied voltage using the wavelength displacement. According to (24), 27.6 V is equivalent to a Bragg shift of 2.48 pm which is very close to the accuracy of Bragg Meters commercially available (approximately ±2.0 pm). Clearly, the final accuracy does not depend only on the measuring capacity of the Bragg Meter, as many other factors interfere, but the Bragg Meter's accuracy imposes a lower limit.

Werneck et al [6] have used a mechanical amplifier in order to multiply the strain and obtain a larger amount of Bragg displacement, that is, a higher sensitivity. The authors used a stack made up of ten PZT rings in a parallel polarization scheme. Also, by decreasing the fiber length in the mechanical setup increases the FBG strain, as will be seen in the sequence.

For calculating the sensitivity of this setup we rewrite (21) including n, the number of PZT elements:

$$\Delta w = n d_{33} V \tag{25}$$

Since the FBG is bonded between the two fixing points of the stack, the displacement previewed by (25) will be transmitted to the fiber, so that

$$\Delta w = \Delta L_{FBG} \qquad (26)$$

Now combining Eq. (5) in [7], (25) and (26), and considering $\Delta T = 0$ (constant temperature environment), we get to:

$$\Delta \lambda_B = \lambda_B (1 - \rho_e) \frac{nd_{33}V}{L_{FBG}} \qquad (27)$$

Substituting the PZT constants in (27) we have the following sensitivity:

$$\frac{\Delta \lambda_B}{\Delta V} = 128.6 \ pm/kV \qquad (28)$$

3.2. Optical setup for DC input

The block diagram of the optical setup for DC input is shown in Figure 13. A variable high voltage DC power supply was used to test the system. An ASE (Amplified Spontaneous Emission) broadband light source illuminates the FBG and its reflected peak is directed to an optical spectrum analyzer (OSA) which is controlled and monitored by a computer, the dotted-line box in Figure 13 encompasses the commercial optical interrogator with a resolution of 2.0 pm. With this setup it is possible to read the wavelength displacement of the sensor and relate it with the DC input voltage applied using a high voltage source. Using this scheme it is possible to test and calibrate the system.

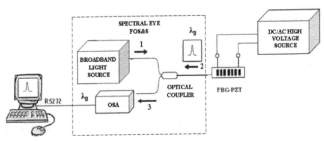

Figure 13. Schematic diagram of the setup for DC input.

3.3. Optical setup for AC input

As mentioned before, commercial OSAs are too slow to respond to the 60-Hz line frequency. For AC voltage measurements we used the interrogation system shown in Figure14. A high voltage AC source was used to supply the input signal to the PZT electrodes. A broadband light source was used to illuminate the FBG-PZT sensor via an optical circulator. The reflected spectrum of the sensor passes through a Fabry-Perot (FP) tunable filter with a 0.89 nm bandwidth. The resulted signal at the output of the FP filter is the convolution between the reflected FBG spectrum and the FP transmission spectrum. The optimum position of the center wavelength of the FP filter is chosen by a novel algorithm developed by [9]. The dashed area on the spectrum drawing at the left of the FP filter in Figure 14 is the intersection between the spectrum of the reflected signal and the band pass of the FP filter. The integral of this area is the total light power that exits the filter and reaches the photodetector. The intersection point of the two spectra occurs at the linear portion of each curve, therefore when the sensor spectrum moves, the superimposed area varies linearly.

Since the sensor spectrum is oscillating at 60 Hz, the intersection will increase and decrease accordingly and the output power of the filter will also oscillate at the same frequency. After this demodulation process, the amplitude (power) of the light signal is proportional to the instantaneous input voltage applied to the PZT. The signal is then fed into an amplified photodetector which output voltage signal is analyzed by an oscilloscope.

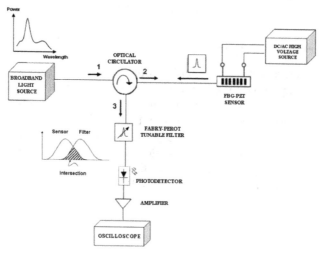

Figure 14. Schematic diagram of the setup for AC voltage using a tunable Fabry-Perot filter.

3.4. Results and discussion

For the first experiment, DC voltages were applied to the FBG-PZT to measure the Bragg displacement by the interrogation system shown in Figure 13. The applied voltage to the PZT ranged from 0 V to 2250 V approximately to not exceed the maximum electric field in direct and reverse directions, specified by the ceramic datasheet. By applying a DC voltage to the PZT and recording the respective Bragg shift, it was possible to plot the graphs shown in Figure 15.

For multiple measurements, we found the average linear sensitivity of 89.09 pm/kV for the FBG-PZT sensor and an average correlation coefficient of $R^2=0.9985$. The recorded sensitivity indicated a Bragg wavelength shift of 89.09 pm per 1000 V applied to the terminals of the PZT.

The voltage dependent sensitivity was calculated and Figure 15 shows the results for 10 measurements. By using a linear fitting procedure(Matlab) the uncertainty studies regarding these results indicate an RMSE of 0.0025 kV and an average standard deviation of 1.19 pm

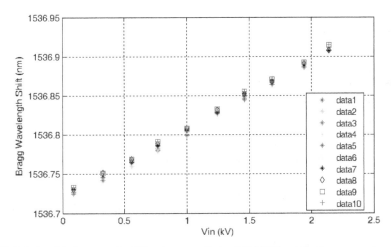

Figure 15. Relationship between the DC voltage applied to FBG-PZT and the Bragg wavelength shift.

which, when divided by the sensitivity, produces an uncertainty in voltage measurement of 13.36 V. This error in voltage represents an uncertainty of 0.09 % in the 13.8 kV Class, which is very close to the 0.2% accuracy needed to comply with the IEC for DC measurements. According to the transducer errors on Figure16, the maximum and minimum residual errors were 0.005 kV and 0.007 kV when measuring0.09 kV and 0.56 kV respectively. This dispersion presented on the results was due to inaccuracies of the system, as well as OSA uncertainty.

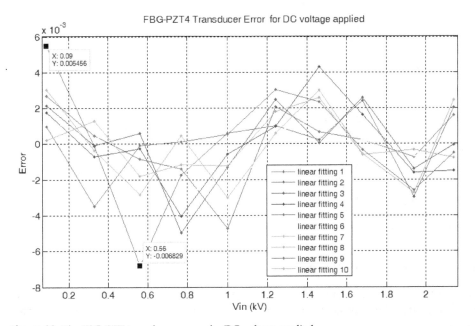

Figure 16. The FBG-PZT transducer errors for DC voltage applied.

The obtained average experimental sensitivity of 89.09 pm/kV for FBG-PZT sensor was smaller than that expected by the theoretical analysis shown in (28): 128.6 pm/kV. The reason for this difference may be related to the elasticity of the materials employed in the mechanical setup. When a voltage is applied to the PZT that forces its displacement, the optical fiber bonded to the aluminum assembly pulls in the opposite direction. If the structure relaxes, the fiber will not be completely stretched, thus preventing full displacement.

The Young's modulus of the optical fiber (70 GPa) is of the same order of magnitude as that of aluminum (69 GPa); however, this cannot be attributed to the aluminum stretching, because its cross section area is much larger than that of the fiber. However, the adhesive used to bond the stack together is fairy elastic, and might easily relax by a few nanometers, which is enough to lower the fiber displacement, resulting in lower sensitivity as the fiber is stressed less than that theoretically calculated.

The AC measurement results consist of measuring the variations in the Bragg wavelength, converted from the photodetector output voltage as a function of AC voltage applied to the terminals of the FBG sensor FBG-PZT. By applying an AC voltage to the PZT and measuring the output voltage of the amplifier over six consecutive cycles, it was possible to obtain the plots shown in Figure 17. The sensor response to an applied voltage in the range of 0 kV to 2 kV, approximately, is presented. A linear relation was found with an average sensitivity of 0.087 and an average correlation coefficient of R^2 0.9983.

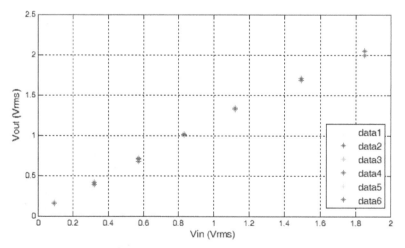

Figure 17. Relationship between the input and output voltages using the Fabry-Perot filter.

Figure 18 shows the linear fitting results for six measurements. According to Figure 18, one can conclude that the maximum error was 0.05 kV on 0.83 kV and the average RMSE was 0.0314 kV.

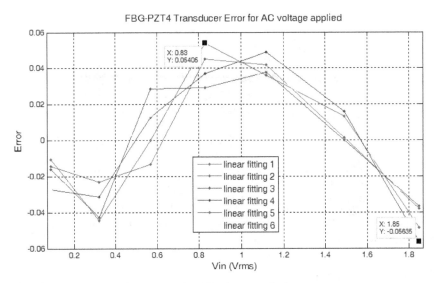

Figure 18. The FBG-PZT transducer errors for AC voltage applied.

For the analysis of dynamic measurements of the output signal it is necessary to measure the total harmonic distortion (THD) of the input signal. Figure 19 shows the wave form of the signal from the measurement system. The objective of this analysis was to investigate the behavior of the input signal (60 Hz line signal) and measure the output harmonic distortion, as the applied AC voltage is increased. The obtained THD was 4.72% and this low value ensures that the harmonic distortion in AC results is not related to the input signal. We can notice that the measurement system output THD occurs particularly when the input voltage reaches the end of the scale, approximately at 1700 V. The reason for this harmonic distortion is that the input signal reaches a nonlinear region of the convolution function between FBG and FP filter spectra. A photograph of the experimental setup can be seen in Figure 20.

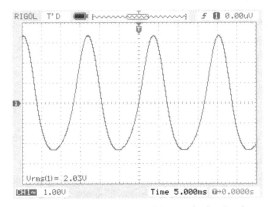

Figure 19. Photodetector output signal for increment of the applied AC voltage (Vrms).

Figure 20. Experiment Setup for AC/DC measurements.

3.5. Temperature drift study

Equation (5) in [7] shows that the temperature also displaces the central wavelength of an FBG. This is because α, the thermal expansion coefficient of the silica, and η, the thermo-optic coefficient, are temperature dependent parameters. However, if the FBG is bonded to an aluminum assembly, the Bragg wavelength will drift with temperature, not only due α and η but also due to the thermal expansion coefficient of the aluminum. Now, if the sensor center wavelength drifts, its gain and sensitivity will vary accordingly.

There are four options to circumvent this effect. The first option is to measure the PZT temperature by using another FBG, and calculating the respective Bragg drift. Hence, an automated controller could adjust the FP filter accordingly. The second option, which only works with twin FBGs, is simply performed by placing the FBG-filter into the same thermal environment as the PZT. In this way, both FBGs will drift together, and the optimum position will be maintained.

The third option requires an engineered assembly so that the thermal expansion is compensated by counter balanced strains. If the PZT stack presented the same strain as the strain of the aluminum ends of the assembly, the fiber and the FBG would not be subject to any strain because the stack displacement would be counterbalanced by the supporting ends. Since the thermal expansion coefficient of aluminum ($24 \times 10^{-6}/°C$) is greater than that of the PZT4 ($4.5 \times 10^{-6}/°C$), we would need a stack length greater than that of the supporting ends to have a counterbalancing effect. Of course, we would also have to include the thermal expansion of the optical fiber ($0.55 \times 10^{-6}/°C$) for the perfect match.

To check this approach the sensor was inserted inside an oven from 25°C to 45°C, without any voltage applied to the PZT4. Figure 21 shows a linear Bragg wavelength displacement with a negative slope as the temperature is varied. The negative slope indicates that the fiber was relaxed instead of being pulled by thermal displacement. The linear fitting for this measurement shows a response for the temperature sensitivity of 55.11 pm/°C with a maximum error of 0.03 kV on 45°C.

In reality, since the thermal coefficient ratio between aluminum and PZT is 5.33, to obtain a perfect match, the total length of aluminum parts should be 5.33 times smaller than that of the PZT stack, which would compromise the project. However a few other materials are available, such as titanium, which has a thermal expansion of $8.6 \times 10^{-6}/°C$. In this case, the ratio of titanium parts over PZT parts would be around 2, which is reasonable.

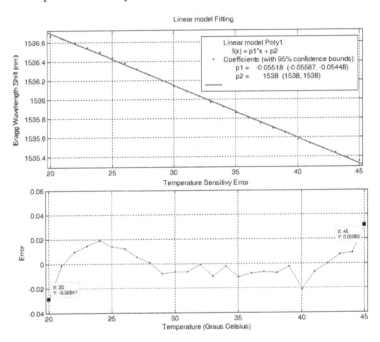

Figure 21. Thermal sensitivity of the FBG-PZT sensor.

3.6. Conclusions

A new optical voltage transformer (OVT) measurement scheme for an AC high voltage line was developed, bearing in mind accuracy, repeatability, and reliability, and was appropriated to in-field operation, and able to comply with the standard IEC 60044-5 [8]. This setup may be used as the core of a practical 13.8 kV-Class VT, as long as temperature compensation is addressed appropriately. As for a possible commercial production of OVTs using the techniques proposed here, it is worth recalling that in recent years the extensive development of optical devices for the telecommunications market has improved their reliability, and, at the same time, decreased their cost. In addition, the devices used in our prototypes, such as PZT crystals, FBG sensors, broadband light source, and an amplified In GaAs photodetector, are not more expensive than the components used in conventional VTs. For this reason, it is considered that the final fabrication costs would not exceed those of conventional VTs.

4. Measuring current in HV transmission lines

4.1. Hybrid devices and magnetostriction

Consistent current measurements are obligatory in several situations, such as in substations, power transmission and oil industry [10, 11]. Usually instrument transformers in electric power facilities are massive and heavy equipment. In this sense, numerous current sensing schemes which make use of the Faraday Effect have been proposed, since optical systems frequently offer a wider dynamic range, lighter weight, improved safety and electromagnetic interference immunity [12].

Several indoor designs of hybrid optical current sensors have been presented, some of which explore the magnetostrictive actuation of a ferromagnetic rod over a fiber Bragg grating, which is attached to this material. All magnetic materials present the magnetostrictive phenomenon, in which the material suffers a strain due to the application of a magnetic field, i. e, the magnetic sample shrinks (negative magnetostriction) or expands (positive magnetostriction) in the direction of magnetization [13]. Thus, magnetostrictive materials (MM) convert magnetic energy into mechanical energy, and the inverse is also true, i.e., when the material suffers an external induced strain its magnetic state is altered. A typical length variation curve showed by this kind of material is presented in Figure 22 [14], where in region 3 the saturation state has been reached ($\Delta L/L$ is the strain). The FBG is frequently attached to the material using a common adhesive, usually cyanoacrylate glue. Thus, as the magnetostrictive rod is strained by the application of a magnetic field (generated, for instance, by an electrical current), also it is the grating, providing a current measurement that is wavelength encoded. This sensor head arrangement is presented in Figure 23.

In [15] the first studies concerning de use of FBGs and MM in DC electrical current measurements are described. In this proposed system two different alloys are employed as the sensor head, being one of them a magnetostrictive rod and both with the same thermal expansion coefficient. In each alloy there is an attached FBG, written in the same optical path. While the Terfenol-D (which is an rare-earth magnetostrictive alloy) rod is used to

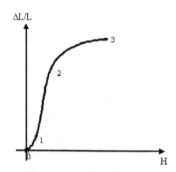

Figure 22. Magnetostrictive material strain characteristic [14].

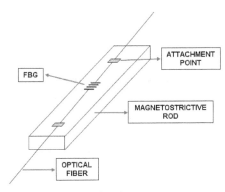

Figure 23. FBG attachment arrangement (sensor head).

obtain the current measurement itself, as it strains the FBG, the grating attached to the non-magnetic alloy provides data concerning temperature variations, acting as a reference for thermal compensations. In this way, when there is a temperature variation both gratings are subjected to the same thermal expansion, the difference between the Bragg wavelengths is a measure of the magnetostriction, and the FBG attached to the non-magnetic alloy (MONEL 400 – a nickel, copper and iron alloy) provides information about the temperature. This sensor head schematic is showed in Figure 24.

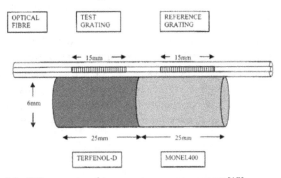

Figure 24. Sensor head for DC-current and temperature measurement [15].

The several advantages showed by the FBG sensing technology in high voltage environments have boosted the development of a number of current measurement systems, especially for the power industry. An optical current transformer (OCT) based on Terfenol-D actuation for AC current signals is proposed in [16], where a prototype to be used in a 100 – 1000 A measurement range was constructed. A testing matrix was made in order to investigate an optimized sensor head operating point, in terms of mechanic and magnetic biases. Since magnestostriction is a unipolar phenomenon, i.e., its output is rectified, to obtain a bipolar response the system must operate in its liner region (region between 1 and 2 in Figure 22) by the application of a DC biasing magnetic field (Figure 25a). This matrix consisted in a C yoke that was turned into a closed loop, the sensor is located in the forth arm (Figure 25b).

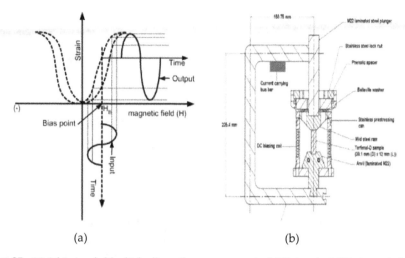

(a)   (b)

Figure 25. (a) A biasing field which allows the measurement of AC signals, (b) Test matrix for system optimization [16].

A monitoring system specifically proposed for transmission lines (TL) is proposed in [17]. In this particular case a different sensor head arrangement is discussed – a coil is not used to excite the optical-magnetostrictive measurement set-up – instead, a uniform FBG with Bragg wavelength of 1531.56 nm is fixed to a magnetostrictive rod, which is positioned at a distance D from the center of the transmission line conductor cable (Figure 26) and parallel to the direction of the current generated magnetic field lines. In this proposal, temperature and TL ac-current can theoretically be measured, where the current is amplitude codified and the temperature is wavelength codified, so there is no ambiguity. The sensor head can be operated at long distances (approximately 2 km) and since an optical fiber path is being employed there is an intrinsic multiplexing capability; however, optical attenuations must be accounted for.

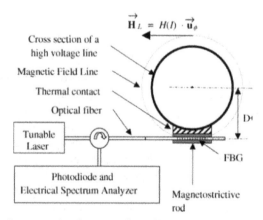

Figure 26. High voltage line sensor head, proposed in [17].

But the usual proposed demodulation techniques are still expensive or not suitable for a prolonged in-field operation. In this section, two opto-mechanical sensor head are showed and compared as current transducers. A laboratory test device for nickel and Terfenol-D is showed, the designed transducers were tested and the temperature effect on the measurement responses was evaluated for a more accurate estimate of the transducers response range.

4.2. Materials and methods

Two transducer configurations are investigated. The main difference lies in the employed materials; nickel and Terfenol-D are utilized as the main parts of the transducer systems. When submitted to a magnetic field generated by the electrical current to be measured, the nickel rod longitudinal length is reduced, while in Terfenol-D the opposite effect occurs, i.e., the material elongates. In this sense, the magnetostrictive samples treatment procedure and the developed electrical current source used to excite the sensor heads are described.

Untreated nickel samples, with dimensions of 100 mm X 100 mm X 10 mm, to be used as magnetic-mechanical transducers need to go through a hot rolling procedure. This method provides thin nickel sheets, thus allowing the reduction of eddy currents. A piece of raw nickel was heated in a metallurgical furnace at a temperature of 600°C for a 10-minute period. After that, the nickel piece is removed from the oven and submitted to metalworking. This procedure was repeated five times, until the nickel sheet reached the thickness of 0.5 mm. This sheet was cut into smaller pieces, with dimensions of 100 mm X 10mm X 0.5 mm, and pilled up in a number of seven, which were glued together with an insulating varnish, composing a complete magnetostrictive part with approximate dimensions of 100 mm X 10 mm X 3.5 mm. This arrangement, besides reducing the effect of eddy currents, is a more robust scheme.

A Terfenol-D rod, an alloy composed by iron, terbium and dysprosium (a Giant Magnetostrictive Material – GMM) with dimensions of 80 mm X 10 mm X 10mm) was used. The material is solid and brittle (Figure 27).

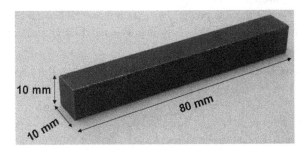

Figure 27. Terfenol-D rod.

Two fiber Bragg gratings were fixed to the MM using commercial cyanoacrylate adhesive, after the cleaning of both nickel and Terfenol-D surfaces, this procedure enables an improved attachment. The FBGs are also stretched, a procedure which will allow the monitoring of magneto-elastic elongations and compressions that the materials may suffer when submitted to magnetic fields. A stretching device is employed in the fibers stretching procedure; while an FBG interrogator was used to monitor the Bragg wavelength shift (Figure 28).

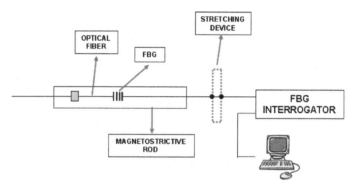

Figure 28. FBG stretching scheme, prior to complete attachment

In order to investigate the magnetostrictive characteristics of both nickel and Terfenol-D samples and to study the proper attachment of both gratings over these rods, a coil was constructed, enabling the excitation of the opto-mechanical transducers with DC current, thus DC magnetic fields. Therefore, the testing system consists in a driving circuit and an exciting coil, fed with DC electrical currents, hence providing a scheme for the magnetostrictive activation.

4.3. Experiments

The transducer devices are positioned in the core of the exciting coil (Figure 29), which is mechanically supported by a PVC tube. Considering this arrangement, a current range of 0 – 27 A_{DC} was delivered to the load, while the Bragg wavelength shift and the transducers temperature were monitored. It is important to monitor the temperature to which the sensor heads are submitted, since there is power dissipation as the DC current is developed in the coil, thus increasing the sensor head temperature. The maximum current was limited by the variable transformer capacity.

The first transducer to be investigated was the one prepared with the stacked nickel sheets, composing a rod. A fiber Bragg grating with λ_B=1538.176 nm at 25°C was attached to the rod, and therefore exposed to the magnetic field generated by the current in the exciting coil. Four measurement cycles were carried out, and the obtained curves are presented in Figure 30. A tendency line that adjusts the data and its equation, the theoretical values of the magnetic field and the temperature of the rod during the experiment, are also presented.

Figure 29. Assembled activation coil, designed according to the procedure described in [18]

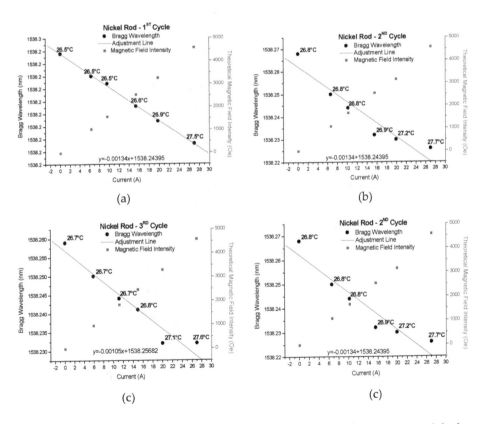

Figure 30. Bragg wavelength X electrical current curves for the FBG-nickel-based sensor head, for four measurement cycles.

The Terfenol-D-based sensor head was also studied. In this case, an FBG with λ_B=1540.065 nm at 25°C was fixed on the alloy rod. Repeating the previously described testing procedure, the obtained Bragg wavelength X electrical current curve is showed in Figure 31. A tendency line that adjusts the data, the theoretical values of the magnetic field, and the temperature of the rod during the experiment, are also presented.

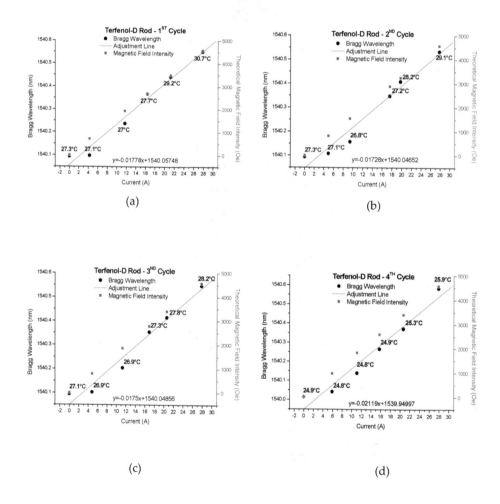

Figure 31. Bragg wavelength X electrical current curves for the FBG-Terfenol-D-based sensor head, for four measurement cycles

4.4. Thermal behavior of sensor

In monitoring applications, in which the strain information is a part of the transduction process, the thermal behavior of the measurement system must be known since most current measurements are done outdoors. This data can be later applied to compensate temperature variation effects. In order to submit the gratings attached to Terfenol-D and nickel rods to a wide temperature range a testing set-up composed by a thermal shaker and a 2000 ml becher with water, where the sensors are immersed (Figure 32), was used.

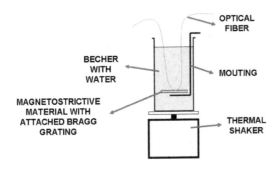

Figure 32. Thermal behavior experiment set-up

Considering that the fixation of the Bragg gratings over the surface of the rods is ideal, the strain over the optical fiber developed during the experiment is due to the magnetostrictive material linear thermal expansion, i.e., ε_m. Thus,

$$\frac{\Delta L}{L_o} = \varepsilon_m = \alpha_M \Delta T \tag{29}$$

where α_M is the MM linear thermal expansion coefficient.

For strain measurements, taking into account the Bragg wavelength expression given by [19]

$$\frac{\Delta \lambda_B}{\lambda_B} = k \cdot \left(\varepsilon_m + \alpha_S \Delta T \right) + \frac{1}{n} \frac{\delta n}{\delta T} \Delta T \tag{30}$$

where $\Delta \lambda_B$ is the Bragg wavelength shift, λ_B is the Bragg wavelength at the beginning of the test, k is the gauge factor (k=0.78), ΔT is the temperature variation, α_S is the silica thermal expansion coefficient, and $1/n \cdot \left(\delta n / \delta T \right)$ is the temperature dependence of the refractive index.

Therefore, using Eq. 29 and 30 one can obtain

$$\frac{\Delta\lambda_B}{\Delta T} = \lambda_B \left(k\alpha_M + k\alpha_S + \frac{1}{n}\frac{\delta n}{\delta T} \right) \qquad (31)$$

The theoretical thermal sensitivity of the opto-mechanical sensor is given by Eq 31. For the nickel rod, with $\lambda_B = 1537.547$ nm (T = 5.6°C) and $\alpha_M(nickel) = 13.3 \times 10^{-6}/°C$, the sensitivity of the Bragg wavelength as a function of temperature is

$$\frac{\Delta\lambda_B}{\Delta T}_{NICKEL} = 0.0258 \quad nm\,/°\,C \qquad (32)$$

Being the magnetostrictive material Terfenol-D, and $\lambda_B = 1539.727$ (T=5.6°C) and $\alpha_M(terfenol)= 12\times10^{-6}/°C$, the sensitivity of the Bragg wavelength as a function of temperature is

$$\frac{\Delta\lambda_B}{\Delta T}_{TERFENOL} = 0.0242 \quad nm\,/°\,C \qquad (33)$$

For the nickel rod, admitting a Bragg wavelength infinitesimal variation for an infinitesimal temperature variation, from Eq. 32 we have

$$d\lambda_B = 0.0258 \cdot 10^{-9}\,dT$$
$$\lambda_B = \int 0.0258 \cdot 10^{-9}\,dT \qquad (34)$$
$$\lambda_B = 0.0258 \cdot 10^{-9}T + C$$

where C is the integration constant for the indefinite integral.

Considering the Bragg wavelength just after the stretching process as the initial condition λ_B=1538.019 nm (T=24.6°C).

Using Eq. 33 one obtain

$$C = 1537.384 \text{ nm.}$$

Hence, the theoretical thermal sensibility for the nickel-based set-up is

$$\lambda_{B(NICKEL)} = 0.0258 \cdot 10^{-9}T + 1537.384\, nm \qquad (35)$$

Repeating this procedure for the Terfenol-D-based set-up, where $\lambda_B = 1540.046$ nm (T = 23.8°C), the obtained value for the integration constant is C = 1539.470 nm. Thus,

$$\lambda_{B(TERFENOL)} = 0.0242 \cdot 10^{-9}T + 1539.470\, nm \qquad (36)$$

In Figures 33 and 34 the measured responses, when both sensor heads are submitted to a temperature variation range of approximately 60°C, are presented. For comparison purposes, a tendency line that adjusts the experimental data and the calculated theoretical thermal sensitivity for each developed set-up are also showed.

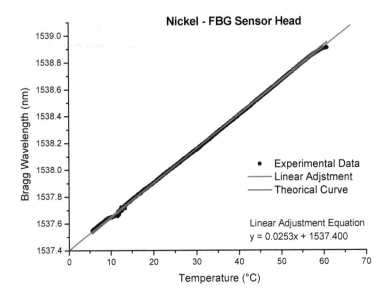

Figure 33. Bragg wavelength as function of temperature for the nickel – FBG sensor head

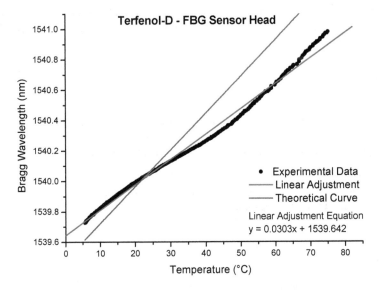

Figure 34. Bragg wavelength as function of temperature for the Terfenol-D – FBG sensor head

4.5. Discussion

The Bragg wavelength for the nickel-based transducer decreases as the current increases, once this material presents a negative magnetostrictive coefficient; whereas the Terfenol-D-based prototype shows a positive magnetostriction (and the Bragg wavelength increases as the current increases).

During the experiments disregarding the temperature effects on the response, the nickel-based transducer showed an average Bragg wavelength shift range of 0.038 nm for a current variation range of 27 A. For the same current range, the Terfenol-D-based transducer showed an average Bragg wavelength shift range of 0.449 nm, approximately one order of magnitude larger than the one showed by the nickel-based transducer. However, considering the temperature effects through the tendency lines obtained during the temperature compensation experiments and showed in Figures33 and 34, this influence has to be taken into account. Thus, in this case, the nickel-based transducer showed an average Bragg wavelength shift range of 0.063 nm, while the Terfenol-D-based transducer exhibited an average Bragg wavelength shift range of 0.417 nm. When thermal effects are compensated the nickel-based transducer Bragg wavelength shift range is increased, since in this particular situation the heating and the magnetostrictive effects work in opposite directions on the FBG response, i.e., while the current induced magnetic field decreases the read Bragg wavelength, the heating increases it.

Nickel and Terfenol-D are suitable materials to be used in current monitoring transducers; yet, the alloy shows an improved response for a specific current range, since it is a giant magnetostrictive material. The development stages of a complete transducer system include the use of greater DC currents to investigate the saturation region of the materials, and the modeling of the response of the transducers when submitted to different mechanical stresses and magnetic biases in order to determine an optimized sensor set-up for both DC and AC current measurements.

Author details

Regina C. S. B. Allil
Instrumentation and Photonics Laboratory, Electrical Engineering Program, Universidade Federal do Rio de Janeiro (UFRJ), RJ, Brazil
Division of Chemical, Biological and Nuclear Defense, Biological Defense Laboratory,
Brazilian Army Technological Center (CTEx) RJ, Brazil

Marcelo M. Werneck, Bessie A. Ribeiro and Fábio V. B. de Nazaré
Instrumentation and Photonics Laboratory, Electrical Engineering Program, Universidade Federal do Rio de Janeiro (UFRJ), RJ, Brazil

5. References

[1] Stone G C, Overview of Hydrogenerator Stator Winding Monitoring, International Conference on Electric Machines and Drives,–IED, May 1999, 806-808, Seattle, WA.

[2] Werneck M.M, Allil R.C.S, Ribeiro B. A, Calibration and operation of a fiber Bragg grating temperature sensing system in a grid-connected hydrogenerator, accepted for publication on IET Science, Measurements and Technology in September 2012.

[3] A. Klimek, Optical Technology: A new Generation of Instrument Transformer, Electricity Today, issue 2, 2003.

[4] S. K. Lee, Electrooptic voltage sensor: birefringence effects and compensation methods, App. Optics,1990, vol. 29, no. 30.

[5] Emerging Technologies Working group and Fiber Optic Sensors Working Group, "Optical Current Transducers for power systems: a review", IEEE Trans. Power Delivery, 9(4), 1778-1788, 1994.

[6] M. M. Werneck, A. C. S. Abrantes, Fiber-optic-based current and voltage measuring system for high-voltage distribution lines, IEEE Transactions on Power Delivery 19 (3): 947-951 Jul 2004.

[7] B. A. Ribeiro, M. M. Werneck, FBG-PZT sensor system for high voltage measurements, 2011 IEEE Instrumentation and Measurement Technology Conference, 10-12, May 2011.

[8] Instrument Transformers-Part 5-Capacitor Voltage Transformers, IEC60044-5, 2004.

[9] B. Ribeiro, M. M. Werneck and J. L. S. Neto, Otimização da demodulação de sinais de um transformador de potencial óptico usando um filtro óptico sintonizável, Congresso Brasileiro de Automática- CBA, Sept. 2012.

[10] D. Reilly, A. J. Willshire, G. Fusiek, P. Niewczas, and J. R. McDonald. A fiber-bragg-grating-based sensor for simultaneous AC current and temperature measurement. IEEE Sensors Journal, 2006, vol. 06,1539–1542.

[11] F. V. B. de Nazaré and M. M. Werneck. Temperature and current monitoring system for transmission lines using power-over-fiber technology, 2010 IEEE Instrumentation and Measurement Technology Conference.779 – 784, May 2010, Austin – Texas - USA.

[12] R. C. S. B. Allil and M. M. Werneck. Optical high-voltage sensor based on fiber Bragg grating and PZT piezoelectric ceramics. IEEE Transactions on Instrumentation and Measurement, vol. 60, n. 6, 2118-2125, June 2011.

[13] E. Hristoforou and A. Ktena. Magnetostriction and magnetostrictive materials for sensing applications. Journal of Magnetism and Magnetic Materials, 372 – 378, 2007.

[14] A. Olabi and A. Grunwald. Design and application of magnetostrictive materials. Materials and Design 2008; 469 – 483.

[15] J. Mora, A. Díez, J. L. Cruz and M. V. Andrés. A magnetostrictive sensor interrogated by fiber gratings for DC-current and temperature discrimination. IEEE Photonics Technology Letters, 1680 – 1682, 2000.

[16] D. Satpathi, J. A. Moore and M. G. Ennis. Design of a Terfenol-D based fiber-optic current transducer. IEEE Sensors Journal, 1057 – 1065, 2005.

[17] J. Mora, Ll. Martínez-León, A. Díez, J. L. Cruz and M. V. Andrés, Simultaneous temperature and ac-current measurements for high voltage lines using fiber Bragg gratings, Sensors and Actuators A 125, 313 – 316, 2005.

[18] G. Engdahl and C. B. Bright, Magnetostrictive Design, in Handbook of Giant Magnetostrictive Materials, G. Engdahl, Academic Press, 207 – 286, 2000.

[19] M. Kreuzer, Strain measurement with fiber Bragg grating sensors, HBM Deutschland, Darmstadt – Germany.

Chapter 8

Fiber Bragg Grating Technology for New Generation Optical Access Systems

Oskars Ozolins, Vjaceslavs Bobrovs, Jurgis Porins and Girts Ivanovs

Additional information is available at the end of the chapter

1. Introduction

Thanks to achievements in photonics technologies optical networks have gained a notably increase in capacity per single fiber during the last decade [1]. Demand for greater transmission speed has been increasing exponentially because of the impulsive spread of Internet services (see Fig. 1) [2]. At the same time, the radical improvement of the capability of digital technologies has made feasible expanding multimedia services [1]. Therefore bandwidth intensive applications and exponential Internet traffic growth are continuing to drive the further penetration of optical fiber into the optical access systems [2].

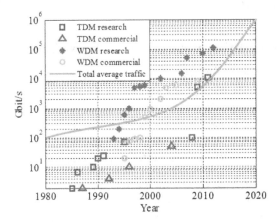

Figure 1. System capacity (per fiber) and network traffic [3]

A new generation access network becomes indispensable to upgrade the systems [4, 5]. The passive optical network (PON) has the feasibility to lead the deployment of new high-capacity and future-proof broadband networks [6]. The use of wavelength division multiplexing (WDM) in the access networks is further defensible in terms of video services support. WDM solutions are forging ahead towards higher data transmission rate and lower channel spacing to utilize available bandwidth more effectively [7].

The main reasons behind the implementation of new generation systems are to meet demand of capacity, user density requirements and scalability, while ensuring that the cost per unit bandwidth is lowest possible [3, 8]. The novel concept is WDM-direct in which multiple wavelengths are directly connected to each optical network unit (ONU) [9, 10]. Increasing spectral efficiency is important for building efficient WDM-direct systems, since this allows the optical infrastructure to be shared among many channels, and thus reduces the cost per transmitted information bit in a fully loaded system [11]. High performance optical filters are groundwork for realization of high speed dense WDM (DWDM)-direct systems where coherent and incoherent crosstalk between adjacent channels becomes a main source of degradation: adjacent channels interfere with each other upon detection, and the resulting beating gives rise to signal distortions, provided that the beat frequencies lie within the bandwidth of the detection electronics [12, 11].

Proposed approach for increasing the transmission capacity is to reduce the channel spacing of a DWDM-direct system to the minimum while keeping the mature and well developed optical filter technologies like fiber Bragg grating (FBG). To realize proposed approach limiting factors must be taken into account. One part of these factors is related to efficient bandwidth of FBG which in proposed approach is determined employing optical signals (transmission speed 2.5 Gbit/s and 10.3125 Gbit/s which conform 2 Gigabit Ethernet (GE) and 10 GE of Ethernet hierarchy) with different wavelength offset value within filter pass-band. Other part of factors is related to evaluation of the minimal channel spacing for concrete FBG in DWDM-direct system.

2. Optimal complex tranfer function for access systems

A FBG is periodic variation of the refractive index along the propagation direction in the core of optical fiber that reflects particular wavelengths of light and transmits all others. Low channel spacing and high data transmission rate sets strict requirements for DWDM filter characteristics and any imperfections in their parameters, such as amplitude and phase responses, becomes critical. Understanding and distracting of those optical filter imperfections to high speed DWDM-direct systems are of great importance [7].

Low channel isolation from adjacent channels is one of these imperfections in optical filter parameters. To ensure high channel isolation we need to inscribe FBG filters with complex apodization profiles. Changes in apodization profiles emerge in different filter bandwidth at -3 dB and -20 dB level and suppression of undesirable side lobes in optical filter amplitude response. Performance of three different apodization profiles and their influence on DWDM-direct systems main parameters: channel spacing and data transmission rate have been evaluated numerically.

2.1. Simulation method

To numerically evaluate impact of different FBG apodization profiles on high speed DWDM-direct system combination of two different simulation programs was used: Bragg Grating Filters Synthesis 2.6 (BGFS 2.6) simulation program for mathematical description of FBG optical filter and OptSim 5.2 simulation program to simulate high speed DWDM-direct systems. In the BGFS 2.6 simulation program different FBG optical filters with defined apodization profiles were realized. This simulation program is based on Transfer Matrix Method (TMM). TMM is used to create a numerical periodic non-uniform FBG filters. It is applied to solve the coupled mode equations and to obtain the spectral response of the fiber Bragg grating. In this approach, the grating is divided into uniform sections. Each section is represented by a 2x2 matrix. By multiplying these matrices, a global matrix that describes the whole grating is obtained (see Fig. 2. and equation 1):

Figure 2. Transfer matrix method used to obtain the spectral characteristics of a fiber Bragg grating (Δn - average index of refraction, Δz - section length, R_0, S_0, R_M, S_M - electromagnetic waves amplitudes)

$$\begin{cases} T_M = T_1 \cdot T_2 \cdot T_3 \cdot \ldots \cdot T_i, \\ \begin{bmatrix} R_M \\ S_M \end{bmatrix} = T_M \cdot \begin{bmatrix} R_0 \\ S_0 \end{bmatrix}. \end{cases} \tag{1}$$

The reflection coefficient of the entire grating is defined as:

$$\rho = \frac{S_0}{R_0}. \tag{2}$$

The main drawback of this method is that M may not be made arbitrarily large, since the coupled-mode theory approximations are not valid when uniform grating section is only a few grating periods long. Thus, it requires $\Delta z \gg T$ [13, 14].

OptSim 5.2 simulation program uses method of calculation that is based on solving a complex set of differential equations, taking into account optical and electrical noise, linear and nonlinear effects. Two ways of calculation are possible: Frequency Domain Split Step (FDSS) and Time Domain Split Step (TDSS) methods. These methods differ in linear operator L calculations: FDSS does it in frequency domain, but TDSS calculates linear operator in the time domain by calculating the convolution product in sampled time. The first method is easy to realize, but it may cause severe errors during simulation. In our simulation we used the second method, TDSS, which despite its complexity grants a precise result. The Split Step method is used in all commercial simulation tools to perform the integration of the fiber propagation equation:

$$\frac{\partial A(t,z)}{\partial z} = \{L + N\} A(t,z), \tag{3}$$

where $A(t,z)$-the optical field; L-linear operator that stands for dispersion and other linear effects; N – operator that is responsible for all nonlinear effects. The idea is to calculate the equation over small spans of fiber Δz by including either linear or nonlinear operator. For instance, on the first span Δz only linear effects are considered, on the second – only nonlinear, on the third – again only linear [15]. Us it is noticed before, in numerical investigation are used two simulation programs: BGFS 2.6 – to realize FBG filters amplitude and phase responses and OptSim 5.2 to numerically evaluate high speed DWDM-direct systems. Realized FBG filter parameters were recorded in data file, which after simple mathematical calculations were used in OptSim 5.2 simulation program to build user defined optical filters.

2.2. Simulation scheme and results

Simulation scheme (see Fig.3.) consists of transmitter, transmission line and receiver. Number of channels is chosen to evaluate influence of nonlinear optical effects (NOE): self – phase modulation (SPM), cross – phase modulation (XPM), four – wave – mixing (FWM) to used optical filters performance.

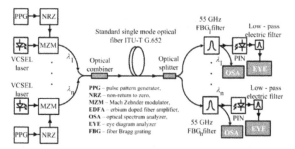

Figure 3. Simulation scheme for DWDM-direct transmission realization with FBG optical filters

The transmitter consists of pseudo-random data source with $2^{31}-1$ bit sequence, non return to zero (NRZ) code former, continuous wavelength (CW) laser source and $LiNbO_3$-based external Mach Zehnder modulator (MZM). The data source produces a pseudo-random electrical signal, which represents the information we want to transmit via optical fibre. Then we use a code former to form NRZ code from incoming pseudo-random bit sequence. The NRZ has long been the dominant code format in fibre optical transmission systems, because of a relatively low electrical bandwidth for the transmitters and receivers and its insensitivity to the laser phase noise [16]. The optical pulses are obtained by modulating CW laser irradiation in MZM with previously mentioned bit sequence. Then formed optical pulses are sent directly to a different length standard single mode fibre (SSMF). The utilized fibre has a large core effective area 80 μm², attenuation $\alpha = 0.2$ dB/km, nonlinear refractive coefficient $n_k = 2.5 \cdot 10^{-20}$ cm/W and dispersion 16 ps/nm/km at the reference wavelength $\lambda = 1550$ nm. Receiver block consists of optical filter, PIN photodiode (typical sensitivity -17 dBm) and Bessel – Thomson electrical filter (4 poles, 7.5 GHz -3dB bandwidth). To simulate insertion loss (polarization dependent loss: 0.1 dB, ripple insertion loss: 0.2 dB, splice and connector loss: 0.1dB) of optical filter we used optical attenuator.

The main idea of our simulations is to demonstrate FBG filters with different apodization profiles (see Fig. 4.) influence on high speed dense WDM communication systems. Investigation of high performance optical band-pass filters are groundwork for realization of high speed dense WDM communication systems.

The main problem is to ensure high channel isolation between adjacent channels. To realize channel isolation performance evaluation of FBG optical filter we used eye diagram, bit error rate (BER) and optical signal spectrum in different system configurations (different channel spacing and data transmission speed). We have chosen three different apodization profiles (see Fig.4.): rectangular, cosinusoidal and Gaussian, four channel spacing values: 200 GHz, 100 GHz, 50 GHz and 25 GHz and two data transmission speeds: 2.5 Gbit/s and 10. 3125 Gbit/s.

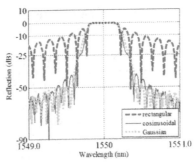

Figure 4. Amplitude response of 55 GHz FBG optical filters with different apodization profiles shown in inset

The results of BER dependence on channel spacing using FBG with rectangular, cosinusoidal and Gaussian apodization profiles are presented in Fig. 5. As we can see systems with 2.5 Gbit/s data transmission speed performance is better (BER values are lower) than systems with 10. 3125 Gbit/s data transmission speed. This can be explained by greater influence of chromatic dispersion on higher data transmission speed optical pulses. In addition, from results we can see that BER values are higher at 25 GHz channel spacing due to greater NOE influence and crosstalk. At both data transmission speeds and all channel spacing values, the worst performance showed the FBG optical filter with rectangular apodization profile. This is mainly because of great undesirable side lobes in optical filter amplitude response. These imperfections in filter amplitude response reduced channel isolation.

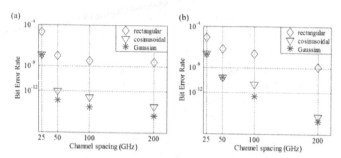

Figure 5. BER dependence on channel spacing using FBG with rectangular, cosinusoidal and Gaussian apodization profiles: a – 2.5 Gbit/s; b – 10.3125 Gbit/s data transmission speed. BER values measured at the worst channel.

As one could see from simulation results the influence of FBG filters with different apodization profiles on high speed DWDM-direct systems is enormous. To ensure high channel isolation and thus realize systems high performance FBG filter must be used with cosinusoidal or Gaussian apodization function because of narrow bandwidth at -20 dB level. FBG with rectangular apodization profile showed the worst performance which resulted in whole system degradation because of imperfections (great side lobes) in amplitude response.

3. Fiber Bragg grating characterization methods

Optical filters in optical transmission systems are a special subgroup of physical components defined in such a way that they select or modify parts of the spectrum of the signal [17, 18]. Signals and physical components can be expressed mathematically by complex functions describing amplitude and phase [17]. Amplitude transfer function describes loss dependency as function of wavelength for optical filter, but phase transfer function is responsible for introduced wavelength dependent amount of delay. There are different methods for transfer funtion characterization. Focus is related to the techniques which are for evaluation of phase transfer funtion and its related parematers.

3.1. Jones matrix eigenanalysis

Polarization mode dispersion (PMD) decreases transmission systems bandwidth and is a fundamental parameter of both: optical fiber and passive optical components. The difference between group delays for two principle states of polarization (PSP) is the differential group delay (DGD) [19]. The PMD value is the average of DGD values. DGD varies randomly with wavelength and time which stands for dispersive effects. Moreover a second-order effect of PMD can lead to optical pulses length changes [20]. As a consequence an optical component with birefringence devoid of chromatic dispersion (CD), can exhibit optical pulses length changes owing to the second-order effects of PMD [21]. PMD is characterized by a Jones Matrix as a function of wavelength [22]. Jones Matrix is common to represent the polarization state of an optical signal or the transfer matrix of a passive optical device. The transfer matrix of an optical device specifies the relationship between the input and output

Jones vectors of the optical signal [20]. This transfer matrix can be characterized by measuring three output Jones vectors in response to three known input Jones vectors. It is worthwhile to note that although it is measured with a few specific polarization states of the optical signal, a Jones Matrix describes a passive optical device such as an optical filter and is independent of the input launching condition of the optical signal [22]. Two basic equations for this DGD estimation are given below. Thus, the two Eigen-values can be calculated from the products of Jones Matrix employing (4).

$$\rho_{1,2} = \frac{m_{11} + m_{22}}{2} \pm \sqrt{(\frac{m_{11} + m_{22}}{2})^2 - m_{11} \cdot m_{22} + m_{12} \cdot m_{21}} \qquad (4)$$

Where $\rho_{1,2}$ is Eigen-values and $m_{11}, m_{12}, m_{21}, m_{22}$ - products of Jones Matrix [22]. Therefore DGD can be expressed as the group delay difference as the function of the optical frequency:

$$\Delta\tau(\omega) = |\tau_{g1} - \tau_{g2}| = \left|\frac{Arg(\rho_1 / \rho_2)}{\Delta\omega}\right| \qquad (5)$$

Where, τ_{g1} and τ_{g2} is group delays for two principle states of polarization, $Arg(\rho_1 / \rho_2)$ denotes the phase angle of ρ_1 / ρ_2 [22]. Equation (5) shows the principle of Jones Matrix technique for DGD measurement. An important practical issue is to choose the frequency step size $\Delta\omega$ for the measurement. For a small step size the measurement would take a long time and instability of devices would have strong impact on the results. For a big step size the output optical signal state of polarization would rotate for more than 45° over each frequency step which leads to inaccurate results [19].

The Jones Matrix technique has a number of advantages compared to other DGD measurement methods [21]. First, it needs only a small wavelength window to perform a measurement. From this point of view method is more suitable for evaluation of detailed wavelength dependency of DGD and PMD. Second, Jones Matrix measurement can be made fast using automated procedures of polarization controller and a polarimeter. Third, the accuracy of the Jones Matrix technique is considered the best compared to other techniques [22]. Therefore Jones matrix Eigen-analysis (JME) is a measurement technique for accurately measure the DGD and PMD of any passive optical component [19].

In this research N7788B component analyser of Agilent Technologies was utilized to perform measurements of FBG with 55 GHz full width half maximum (FWHM) bandwidth. This technology is based on the JME which is the standard method for measuring DGD and PMD of passive optical devices [23].

Measurement scheme (see Fig.6) consists of LiNbO₃ polarization controller, polarimeter, tunable laser source and device under test (DUT): FBG with 55 GHz FWHM bandwidth.

Fig.7 shows measured amplitude transfer function and DGD as function of wavelength for FBG with 55 GHz FWHM bandwidth. Results show that DGD value within filters pass-band does not exceed 70 picoseconds.

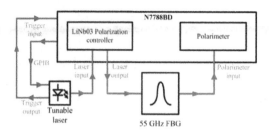

Figure 6. DGD measurement scheme [24]

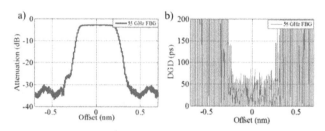

Figure 7. Measured amplitude transfer (a) function and differential group delay (b) of 55 GHz FBG

DGD and PMD value increases at the edges of the optical filter amplitude transfer function which could be a degradation factor for optical signal transmitted through device [25]. Therefore the relative bandwidth available to each broadband access systems channel is reduced, meaning that the channel experiences the effect of the edge of the pass-band of the filter transfer function, where the dispersion effects is expected to be most significant [17].

3.2. Modulation phase shift method

Agilent Technologies 86038B photonic dispersion and loss analyzer was employed for 55 GHz FBG filter parameters evaluation and numerous parameters were obtained: attenuation, group delay (GD) and chromatic dispersion as functions of wavelength. Test equipment is based on the modulation phase shift (MPS) method. In the conventional MPS method, light from a sinusoidal source is intensity modulated before being launched into the device under test [24]. MPS method obtains the group delay response of a device under test by measuring the change in phase of a sinusoidal radio frequency (RF) modulation envelope as the wavelength is changed [23].

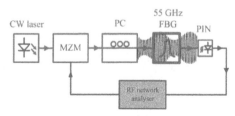

Figure 8. MPS method scheme for 55 GHz FBG measurements

See MPS method realization in Fig.8. The optical light source is a tunable distributed feedback laser. Light from a laser is sent to external MZM and is amplitude modulated (typically in the 100 MHz to 1.25 GHz range). After propagating through the DUT, the transmitted signal is detected by a PIN photodiode. A RF network analyzer is employed to provide a modulating signal of frequency f_m and to measure electrical phase difference between input and output signals [26, 27]. In practice, the wavelength is swept and the change in the group delay $\Delta\tau$ for each wavelength increment is calculated from the measured change in phase according to (6):

$$\Delta\tau_{(\Delta\lambda)} = -\frac{\Delta\varphi}{360^0} \cdot \frac{1}{f_m}. \tag{6}$$

Where the first factor is defined as the fractional cycle of RF phase shift and the second factor represents the period of the RF signal. The subscript $\Delta\lambda$ indicates that the change in group delay being measured was produced in response to an incremental change in wavelength. In (6) we can notice how the group delay and the measured electrical phase present opposite slopes [20, 27].

The attribute called dispersion is defined by:

$$D = \frac{\Delta\tau}{\Delta\lambda} \tag{7}$$

Where $\Delta\tau$ is the change in group delay in seconds corresponding to a change in wavelength $\Delta\lambda$ in meters. In real world applications, the dispersion parameter is given in units of picoseconds per nanometer. Combining (6) and (7), we obtain:

$$\Delta\varphi = -360^0 \cdot D \cdot \Delta\lambda \cdot f_m \tag{8}$$

Equation (8) specifies that the amount of phase change obtained in response to a wavelength step is the product of total device dispersion, wavelength step and modulation frequency. This equation provides several key insights into the capabilities of the MPS measurement method. In order to achieve accurate measures it is important to have a stable wavelength step size, which completely depends on the tunable laser stability [20, 27].

Fig. 9 shows measured attenuation, GD and CD as function of wavelength for FBG with 55 GHz FWHM bandwidth. The insertion loss is 5,3 dB while its bandwidth at -1 dB level is 50 GHz and its bandwidth at -20 dB level is equal to 75 GHz. The group delay variation is limited to 50 ps in the pass-band and the dispersion at the center wavelength is equal to 0 ps/nm. The maximum dispersion in the bandwidth at -3 dB level is found to be within the range of -500 ps/nm to 500 ps/nm.

3.3. Efficient bandwidth measurement method

The main idea of our experiments is to evaluate efficient bandwidth of FBG with 55 GHz FWHM bandwidth. Efficient bandwidth of passive device provides limitations which are required to take into consideration for realization of DWDM-direct transmission system for broadband access.

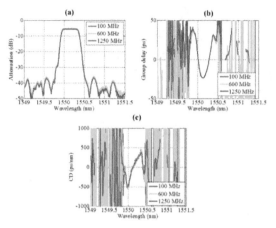

Figure 9. Measured attenuation (a), group delay (b), chromatic dispersion (c) and differential group delay (d) as function of wavelength for 55 GHz FBG filter

3.3.1. Method setup

Efficient bandwidth measurement scheme (see Fig. 10) was realized to investigate FBG with 55 GHz FWHM bandwidth with 2 GE and 10 GE optical signals. The efficient bandwidth measurement scheme consists of typical optical transmission system elements. The transmitter consists of pseudo-random data generator with 2^{31}-1 bit sequence, non-return to zero code former, continuous wavelength laser source and LiNbO$_3$-based external Mach Zehnder modulator.

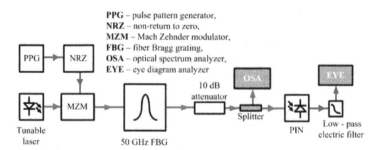

Figure 10. Efficient bandwidth measurement scheme

The data source produces a pseudo-random electrical signal, which represents the information. Then a code former is used to form NRZ code from incoming pseudo-random bit sequence. The optical pulses are obtained by modulating CW laser irradiation in MZM. Formed optical pulses are sent directly to a DUT at different CW laser central wavelength offset values. Receiver block consists of optical attenuator, PIN photodiode and Bessel –

Thomson electrical filter (4 poles, 7.5 GHz -3dB bandwidth). Attenuator with 10 dB rated value was used to simulate loss of 20 km optical fiber, splicing and connectors in direct access systems. Oscilloscope and optical spectrum analyser (OSA) was used to perform measurements of eye diagram and optical power spectral densities, accordingly.

3.3.2. Results for 2.5 GE and 10 GE transmission speed

The bit error rate measurement is a simple method for systems performance evaluation. The error counting in a practical system for realistically low BER values (< 10^{-12}) can be a long process. Therefore the International Telecommunications union (ITU) has created the eye diagram masks for different bit rates with a definite BER value [28].

Fig. 11 shows the eye diagrams and optical power spectral densities of 2 GE optical signals after FBG with 55 GHz FWHM bandwidth for different laser central wavelength offset values (-0.2 nm, -0.1 nm, 0 nm, 0.1 nm, 0.2 nm). Offset value was changed within FBG device pass-band with 0.1 nm step. This value was chosen to fit DWDM systems wavelength grid defined in ITU-T G.694.1 recommendation. As we can see from results greater optical signal amplitude and phase distortions are at the edges of the band-pass optical filter. On Fig.11.a and Fig.11.e are shown eye diagrams for -0.2 nm and +0.2 nm offset values and there are signal waveform degradation. From these results FBG efficient bandwidth is 0.4 nm or 50 GHz and is the same as FWHM bandwidth.

Fig. 12 depicts out the eye diagrams and optical power spectral densities of 10 GE optical signals after FBG with 55 GHz FWHM bandwidth for different laser central wavelength offset values (same as in Fig. 11). On Fig.5.a and Fig.5.e are shown eye diagrams for -0.2 nm and +0.2 nm offset values and there are signal and mask crossing which means that defined BER value is exceeded. The results show that efficient bandwidth is 0.2 nm or 25 GHz and is 0.2 nm or 25 GHz lower than FWHM bandwidth for this transmission speed.

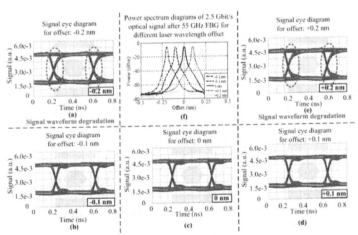

Figure 11. Eye diagrams (a-e) and optical power spectral densities (f) of 2GE optical signal after FBG with 55 GHz FWHM bandwidth for different CW laser wavelength offset shown in inset

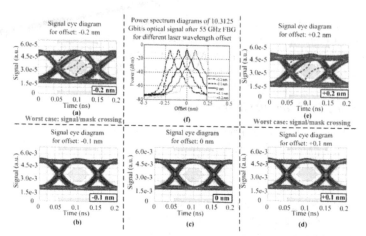

Figure 12. Eye diagrams (a-e) and optical power spectral densities (f) of 10GE optical signal after FBG with 55 GHz FWHM bandwidth for different CW laser wavelength offset shown in inset

4. Realization of new generation access system

Optical band-pass filter performance must be evaluated to increase spectral efficiency of overall optical transmission system [8]. Due to this, detailed investigation has been done into the FBG filter influence on the optical signals in DWDM-direct. For this purpose we have created the DWDM-direct measurement scheme and determined minimal channel interval for 55 GHz FBG filter at which the bit error ratio is sufficiently low. This evaluation was carried out employing eye diagrams and optical power spectral densities of the received optical signal.

4.1. Measurement setup

DWDM-direct scheme (see Fig. 13) is composed of three parts: a transmitter, an optical fiber, and a receiver. The transmitter consists of a pseudo-random data source with 2^{31}-1 bit sequence (Anritsu MU181020A), a non-return-to-zero code former (Anritsu MU181020A), a tunable continuous wavelength laser source (Agilent 81989A, 81949A), and an Avanex LiNbO3-based external MZM. The data source generates a pseudo-random electrical signal which contains the information to be transmitted via optical fiber. Then a code former is used to form an NRZ code from the incoming pseudo-random bit sequence. The optical pulses are obtained by modulating CW laser light in MZM with the generated bit sequence. After optical modulation the formed optical pulses are sent directly to a 20 km SSMF (G.652.d). The utilized fiber has a large core effective area of 80 μm^2, attenuation $\alpha = 0.2$ dB/km, nonlinear refractive coefficient $n_k = 2.5 \cdot 10^{-20}$ cm/W, and dispersion 16 ps/nm/km at the reference wavelength $\lambda = 1550$ nm. The receiver block consists of an optical filter (55 GHz FBG), a PIN photodiode, and a Bessel–Thomson's electrical filter (4 poles, 7.5 GHz -3dB bandwidth, Anritsu MP1026A) [16].

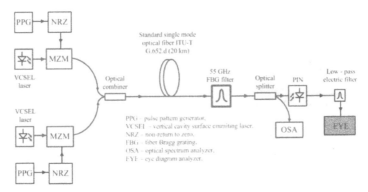

Figure 13. Realized DWDM– direct system for broadband access

In this research the minimal channel interval for DWDM-direct systems with 55 GHz FBG filter has been determined employing measured eye diagrams an optical power spectrum densities. A high-frequency oscilloscope Anritsu MP1026A was used to perform the eye diagram measurements and the optical spectrum analyzer ADVANTEST Q8384 was employed to get optical power spectral densities.

4.2. Spectral efficiency enlargement

Fig. 14 shows the eye diagrams and optical power spectral densities of a 2.5 Gbit/s DWDM-direct system realized with 55 GHz FBG after 20 km of SSMF for different channel intervals from 25 GHz to 125 GHz with 25 GHz (0.2 nm in a wavelength range) step. The step value was chosen to fit DWDM wavelength grid defined in ITU-T G.694.1 recommendation. Both signal detection in the 2.5 Gbit/s DWDM system, 55 GHz FBG, was observed with a 25 GHz channel interval. To reduce undesirable adjacent signal, the channel interval was increased, which gave lower BER values for the detected signal. As a result, the adjacent channel was suppressed more efficiently, because the steepness of a 55 GHz FBG device is very good and the adjacent channel's isolation is ~35 dB. As one can see from the results (Fig. 14b), a 50 GHz channel interval is sufficient to ensure the appropriate BER value for adequate system's performance. The results for greater channel intervals (75 GHz, 100 GHz and 125 GHz, Fig. 14c–e) are also shown to demonstrate DWDM-direct system's stability in the employed spectral range.

The eye diagrams and optical power spectral densities of a 10 Gbit/s DWDM-direct system for broadband access after 20 km of SSMF for the same channel intervals as in the previous case are shown in Fig. 15. Due to a higher transmission speed, the optical power spectral density is broader, which results in stronger influence of CD on the signal quality. This leads to greater degradation of the optical signal, which emerges as a larger standard deviation and jitter for "0" and "1" levels in eye diagram. Similar to the above, a 50 GHz channel interval is sufficient to ensure the appropriate BER value for normal performance of the system at 10 Gbit/s transmission speed; the spectral efficiency is in this case improved from 0.18 bit/s/Hz to 0.2 bit/s/Hz.

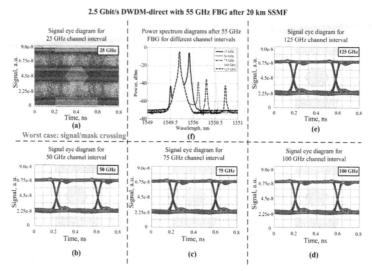

Figure 14. Eye diagrams (a–e) and optical power spectral densities (f) of 2.5 Gbit/s DWDM-direct system realized with a 55 GHz FBG after 20 km of SSMF for different channel intervals (shown in insets).

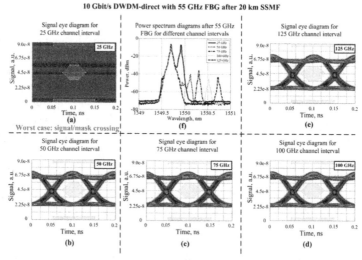

Figure 15. Eye diagrams (a–e) and optical power spectral densities (f) of 10 Gbit/s DWDM-direct system realized with a 55 GHz FBG after 20 km of SSMF for different channel intervals (shown in insets).

4.3. Channel number enlargement: numerical evaluation

Simulation scheme is shown in Fig. 3. Channel count of simulation scheme depends on simulation setup. Two, four and eight channels were chosen balancing between total capacity on one hand and physical limitations on the other.

The main idea of simulations is to demonstrate the possibility of channel number enlargement for FBG filter with 55 GHz FWHM bandwidth in DWDM-direct transmission system for broadband access.

Fig.16.a-c. depicts out power spectral densities and eye diagrams of 2.5 Gbit/s DWDM-direct transmission system with different channel count after 20 km of SSMF and Fig.16.d. shows BER dependence on distance for 50 GHz FBG. We can see that adjacent channel isolation value (~ 30 dB) for 55 GHz FBG is sufficient to realize reliable transmission at eight channel case. In this case influence of adjacent channel caused impairments is minimized by proper optical band-pass filter parameter selection.

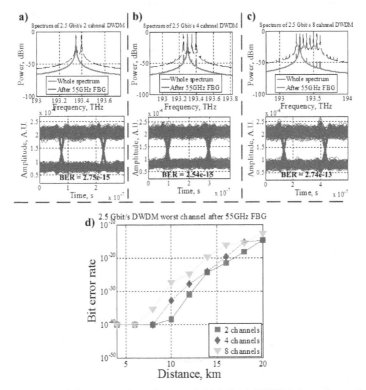

Figure 16. Power spectral densities and eye diagrams of 2.5 Gbit/s DWDM-direct a) two channels, b) four channels, c) eight channels system after 20 km of SSMF and d) BER vs. Distance with 55 GHz FBG optical filter. Results obtained at the worst channel.

Fig.17.a-c. depicts out power spectral densities and eye diagrams of 10 Gbit/s DWDM-direct transmission system with different channel count after 10 km of SSMF and Fig.17.d. shows BER dependence on distance for 55 GHz FBG. Transmission at higher data speed is more affected by chromatic dispersion of optical fibre and total power budget of system is reduced because of greater excess loss in MZM and lower receiver sensitivity for appropriate BER threshold.

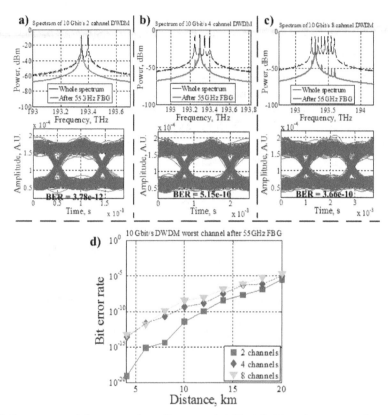

Figure 17. Power spectral densities and eye diagrams of 10 Gbit/s DWDM-direct a) two channels, b) four channels, c) eight channels system after 10 km of SSMF and d) BER vs. Distance with 55 GHz FBG optical filter. Results obtained at the worst channel.

5. Conclusions

As we can see from the results, the proper selection of optical filter amplitude transfer functions is of great importance. In this investigation influence of adjacent channel caused impairments is minimized by proper optical band-pass filter parameter selection. Reliable transmission is realized for 2.5 Gbit/s and 10 Gbit/s DWDM-direct with 55 GHz FBG for 20 km of SSMF.

Results show that DGD value within filters pass-band does not exceed 70 picoseconds for FBG with 50 GHz FWHM bandwidth. DGD value increases at the edges of the optical filter amplitude transfer function which could be a degradation factor for optical signal transmitted through device. Furthermore efficient bandwidth was evaluated for FBG devices employing optical signals with different transmission speed. Efficient bandwidth of 55 GHz FBG device was 0.4 nm or 50 GHz for 2GE optical signal and 0.2 nm or 25 GHz for 10GE optical signal.

We have realized a DWDM-direct system for broadband access that includes FBG filter with 55 GHz FWHM bandwidth. From the measurement results we found the minimal channel interval for the 55 GHz FBG to ensure reliable data transmission, and therefore were able to increase the spectral efficiency of the whole DWDM-direct system for broadband access. In 2.5 Gbit/s and 10 Gbit/s DWDM-direct systems with 55 GHz FBG the detection of both signals were observed for a 25 GHz channel interval. To achieve single-channel detection and suppression of the adjacent channel's power level the channel spacing was increased to 50 GHz. As a result the spectral efficiency of the 10 Gbit/s DWDM system with 55 GHz FBG was raised from 0.18 bit/s/Hz to 0.2 bit/s/Hz.

Author details

Oskars Ozolins, Vjaceslavs Bobrovs, Jurgis Porins and Girts Ivanovs
Riga Technical University, Telecommunications Institute, Latvia

Acknowledgment

This work has been supported by the European Regional Development Fund in Latvia within the project Nr. 2010/0270/2DP/2.1.1.1.0/10/APIA/VIAA/002.

6. References

[1] Kazovsky L. G., Shaw W.-T., Gutierrez D., Cheng N., and Wong S.-W. Next-Generation Optical Access Networks. Journal of Lightwave Technology 2007; 25(11) 3428-3442.

[2] Wong E. Next-Generation Broadband Access Networks and Technologies. Journal Of Lightwave Technology 2012 30(4) 597-608.

[3] Hecht J. Recycled Fiber Optics. How Old Ideas Drove New Technology. Optics and Photonics News 2012; 23(2) 22-29.

[4] Kataoka N., Wada N., Xu W., Cincotti G. and Kitayama K.-I. 10Gbps-Class, bandwidth-symmetric, OCDM-PON system using hybrid multi-port and SSFBG en/decoder. 14th Conference on Optical Network Design and Modeling (ONDM) 2010, March 15 2010.

[5] Effenberger F. J., Kani J., and Maeda Y. Standardization Trends and Prospective Views on the Next Generation of Broadband Optical Access Systems. IEEE Journal on Selected Areas in Communications 2010; 28(6) 773-780.

[6] Kehayas E. Designing Wavelength-Division-Multiplexed Optical Access Networks Using Reflective Photonic Components. 14th Conference on Optical Network Design and Modeling (ONDM) 2010, March 15 2010.

[7] Agrawal G. Nonlinear Fiber Optics (Third Edition). USA: Academic Press; 2001.

[8] Segarra J., Sales V. and Prat J. Agile Reconfigurable and Traffic Adapted All-Optical Access-Metro Networks" 11th International Conference on Transparent Optical Networks (ICTON) 2009, June 28 2009.

[9] Ozolins O., Bobrovs V., Ivanovs G. Efficient Bandwidth of 50 GHz Fiber Bragg Grating for New-Generation Optical Access. 19th Telecommunications forum TELFOR 2011, November 22-24, 2011.

[10] Miyazawa T., Harai H. Optical access architecture designs based on WDM-direct toward new generation networks. IEICE Transactions on Communications 2010; E93-B(2) 236-245.

[11] Pfennigbauer M., Winzer P. J. Choice of MUX/DEMUX filter characteristics for NRZ, RZ, and CSRZ DWDM systems. Journal of Lightwave Technology 2006; 24(4) 1689 – 1696.

[12] Shen Y., Lu K., and Gu W., Coherent and Incoherent Crosstalk in WDM Optical Networks Journal Of Lightwave Technology1999; 17(5) 759-764.

[13] Phing H.S., Ali J., Rahman R.A. and Tahir B.A. Fiber Bragg grating modeling, simulation and characteristics with different grating lengths. Journal of Fundamental Sciences 2007; 3(1) 167-175.

[14] Kashyap R. Fiber Bragg grating. USA: Academic Press; 1999.

[15] Bobrovs V., Ivanovs G. Parameter Evaluation of a Dense Optical Network. Electronics and Electrical Engineering 2006; 4(25) 33-37.

[16] Ozolins O., Bobrovs V., Ivanovs G. Efficient Wavelength Filters for DWDM Systems. Latvian Journal of Physics and Technical Sciences 2010; 6(1) 13–24.

[17] Venghaus H. Wavelength filters in fibre optics. Berlin: Springer; 2006.

[18] Azadeh M. Fiber Optic engineering. London: Springer; 2009.

[19] Chen L., Zhang Z., Bao X. Polarization dependent loss vector measurement in a system with polarization mode dispersion. Optical Fiber Technology 2006 12(1) 251–254.

[20] Hui R., O'Sullivan M. Fiber optic measurement techniques. Burlington: Elsevier; 2009.

[21] Haro J., Horche P.R. Evolution of PMD with the temperature on installed fiber Optical Fiber Technology 2008; 14(3) 203–213.

[22] Yao X. S., Chen X., Liu T. High accuracy polarization measurements using binary polarization rotators. Optics Express 2010; 18(7) 6667-6685.

[23] Kun X., Dai Y., Jin M., Jia F., Chen X., Li X., Xie S. A novel method of automatic polarization measurement and its application to the higher-order PMD measurement. Optics Communications 2003; 215(12) 309–314.

[24] Agilent Technologies. Agilent N7788B/BD optical component analyser. Agilent Technologies, Inc. 2008;11-6.

[25] Westhauser M., Finkenbusch M., Remmersmann C., Pachnicke S., Krummrich P. M. Optical Filter-Based Mitigation of Group Delay Ripple- and PMD- Related Penalties for High Capacity Metro Networks. Journal of Lightwave Technology 2011; 29(16) 2350 - 2357.

[26] Peucheret C. Fibre and component induced limitations in high capacity optical networks. PhD thesis, Technical University of Denmark; 2004.

[27] Agilent Technologies. Agilent 86038B Photonic Dispersion and Loss Analyzer, 2nd ed. Germany: Agilent Technologies Manufacturing GmbH; 2006.

[28] Ozolins O., Bobrovs V., Ivanovs G. DWDM Transmission Based on the Thin-Film Filter Technology. Latvian Journal of Physics and Technical Sciences 2011 3(5) 55–65.

Permissions

The contributors of this book come from diverse backgrounds, making this book a truly international effort. This book will bring forth new frontiers with its revolutionizing research information and detailed analysis of the nascent developments around the world.

We would like to thank Christian Cuadrado-Laborde, for lending his expertise to make the book truly unique. He has played a crucial role in the development of this book. Without his invaluable contribution this book wouldn't have been possible. He has made vital efforts to compile up to date information on the varied aspects of this subject to make this book a valuable addition to the collection of many professionals and students.

This book was conceptualized with the vision of imparting up-to-date information and advanced data in this field. To ensure the same, a matchless editorial board was set up. Every individual on the board went through rigorous rounds of assessment to prove their worth. After which they invested a large part of their time researching and compiling the most relevant data for our readers. Conferences and sessions were held from time to time between the editorial board and the contributing authors to present the data in the most comprehensible form. The editorial team has worked tirelessly to provide valuable and valid information to help people across the globe.

Every chapter published in this book has been scrutinized by our experts. Their significance has been extensively debated. The topics covered herein carry significant findings which will fuel the growth of the discipline. They may even be implemented as practical applications or may be referred to as a beginning point for another development. Chapters in this book were first published by InTech; hereby published with permission under the Creative Commons Attribution License or equivalent.

The editorial board has been involved in producing this book since its inception. They have spent rigorous hours researching and exploring the diverse topics which have resulted in the successful publishing of this book. They have passed on their knowledge of decades through this book. To expedite this challenging task, the publisher supported the team at every step. A small team of assistant editors was also appointed to further simplify the editing procedure and attain best results for the readers.

Our editorial team has been hand-picked from every corner of the world. Their multi-ethnicity adds dynamic inputs to the discussions which result in innovative outcomes. These outcomes are then further discussed with the researchers and contributors who give their valuable feedback and opinion regarding the same. The feedback is then collaborated with the researches and they are edited in a comprehensive manner to aid the understanding of the subject.

Apart from the editorial board, the designing team has also invested a significant amount of their time in understanding the subject and creating the most relevant covers. They scrutinized every image to scout for the most suitable representation of the subject and create an appropriate cover for the book.

The publishing team has been involved in this book since its early stages. They were actively engaged in every process, be it collecting the data, connecting with the contributors or procuring relevant information. The team has been an ardent support to the editorial, designing and production team. Their endless efforts to recruit the best for this project, has resulted in the accomplishment of this book. They are a veteran in the field of academics and their pool of knowledge is as vast as their experience in printing. Their expertise and guidance has proved useful at every step. Their uncompromising quality standards have made this book an exceptional effort. Their encouragement from time to time has been an inspiration for everyone.

The publisher and the editorial board hope that this book will prove to be a valuable piece of knowledge for researchers, students, practitioners and scholars across the globe.

List of Contributors

Marcelo M. Werneck, Bessie A. Ribeiro and Fábio V. B. de Nazaré
Instrumentation and Photonics Laboratory, Electrical Engineering Program, Universidade Federal do Rio de Janeiro (UFRJ), RJ, Brazil

Regina C. S. B. Allil
Instrumentation and Photonics Laboratory, Electrical Engineering Program, Universidade Federal do Rio de Janeiro (UFRJ), RJ, Brazil
Division of Chemical, Biological and Nuclear Defense, Biological Defense Laboratory, Brazilian Army Technological Center (CTEx) RJ, Brazil

Jonghun Lee and Cherl-Hee Lee
Robotics Research Division, Daegu Gyeongbuk Institute of Science & Technology, Daegy, Korea

Kwang Taek Kim
Department of Optoelectronics, Honam University, Gwangju, Korea

Jaehee Park
Department of Electronic Engineering, Keimyung University, Daegu, Korea

Fei Xu, Jun-Long Kou and Yan-Qing Lu
College of Engineering and Applied Sciences and National Laboratory of Solid State Microstructures, Nanjing University, Nanjing, P. R. China

Ming Ding and Gilberto Brambilla
Optoelectronics Research Centre, University of Southampton, Southampton, SO17 1BJ, United Kingdom

David Krčmařík
Institute of Photonics and Electronics AV CR, v.v.i, Praha, Czech Republic

Mykola Kulishov
HTA Photomask, 1605 Remuda Lane San Jose, CA, USA

Radan Slavík
Institute of Photonics and Electronics AV CR, v.v.i, Praha, Czech Republic
Optoelectronics Research Centre, University of Southampton, Southampton, Great Britain

Sergiy Korposh and Seung-Woo Lee
Graduate School of Environmental Engineering, the University of Kitakyushu, 1-1 Hibikino, Wakamatsu-ku, Kitakyushu, Japan

Stephen James and Ralph Tatam
School of Engineering, Cranfield University, Cranfield, Bedford, UK

Alexandre Ferreira da Silva
MIT Portugal Program, School of Engineering, University of Minho, Guimarães, Portugal

Rui Pedro Rocha, João Paulo Carmo and José Higino Correia
Department of Industrial Electronics, University of Minho, Guimarães, Portugal

Oskars Ozolins, Vjaceslavs Bobrovs, Jurgis Porins and Girts Ivanovs
Riga Technical University, Telecommunications Institute, Latvia